FOURTH EDITION

The Naval Institute
Guide to
Naval Writing

C. E. Crane and Robert Shenk

Naval Institute Press
Annapolis, Maryland

Naval Institute Press
291 Wood Road
Annapolis, MD 21402

Library of Congress Cataloging-in-Publication Data
Names: Crane, C. E. (Christopher E.), author. | Shenk, Robert, author. | United States Naval
   Institute.
Title: The Naval Institute guide to naval writing / C. E. Crane and Robert Shenk.
Other titles: Guide to naval writing
Description: Fourth edition. | Annapolis, Maryland : Naval Institute Press, [2024] |
   Series: Blue & gold professional library | Includes bibliographical references and index.
Identifiers: LCCN 2023058163 (print) | LCCN 2023058164 (ebook) |
   ISBN 9781682476154 (paperback) | ISBN 9781682479216 (ebook)
Subjects: LCSH: United States. Navy—Records and correspondence. | United States.
   Marine Corps—Records and correspondence. | Naval art and science—Authorship. |
   Technical writing—United States. | BISAC: TECHNOLOGY & ENGINEERING /
   Technical Writing | REFERENCE / General
Classification: LCC VB255 .S54 2024 (print) | LCC VB255 (ebook) |
   DDC 808.06/6359—dc23/eng/20240206
LC record available at https://lccn.loc.gov/2023058163
LC ebook record available at https://lccn.loc.gov/2023058164

♾ Print editions meet the requirements of ANSI/NISO z39.48–1992 (Permanence of Paper).
Printed in the United States of America.

32 31 30 29 28 27 26 25 24   9 8 7 6 5 4 3 2 1
First printing

Figures created by authors unless otherwise noted.

≈

To Sonmin, Elliott, Christopher, Gabriel, and Olivia.
Whether you wear a uniform or not, may your writing
be the fruit of clear thinking and right reason.

≈

Contents

PREFACE

Robert Shenk's third edition (2008) and the prior editions are masterful assemblies of practical, relatable, and readable advice for successful writing in the Navy and Marine Corps. So why do we need a new edition?

Although the general principles of effective writing do not change much from year to year, language usage and communication styles in the culture as a whole do change slightly, and reader preferences change as styles go in and out of fashion. Therefore, some of the techniques of explaining and demonstrating those principles can benefit from updates in the presentation.

In addition, as culture and technology change, the contexts for writing in the Navy sometimes require updating the guidance offered to Sailors and Marines. For example, the use of email has expanded even beyond where it was in 2008, and social-media writing was not a part of official Navy communications at all at that time.

WHAT'S NEW?

This edition offers expanded advice in several areas. Chapter 2 now incorporates the previous edition's email chapter with letters and memos to better highlight email as a vital form of Navy correspondence and provides expanded guidance on writing email messages.

The previous "Staff Writing" chapter is now "Staff Writing and Operational Documents" (chapter 3), with an increased focus on document types consistent with the other chapters in this edition. The chapter on evaluations and fitness reports incorporates some tips and examples based more on current best practices for what promotion boards are looking for; it also includes expanded advice on Marine Corps evaluations. This edition also now provides general guidance on writing social media posts for your command in chapter 10, which is updated slightly to reflect changes in publications and current practice.

The Plain Writing Act of 2010 has become law since the last edition. I (Crane) have incorporated additional guidance on "plain language" (usually referred to as "plain English" in prior editions) that apply to all federal agencies. Each chapter now includes an executive summary to promote greater structural clarity and to help you find what you need. Finally, many explanations, guidance, and examples are now adapted or updated. All chapters are reorganized somewhat with some new material and revisions to style, but they retain their original and excellent advice on different forms of writing.

During final preparations of this new edition, ChatGPT and other artificial intelligence (AI) tools to produce written texts have become more widespread, raising many questions about the future of writing. Although little has yet changed officially in naval writing, the landscape (or seascape) may look very different in a few years. Yet the likelihood that many writers may begin using AI to generate text in professional settings *increases*, not decreases, their need to have a grasp of the fundamental principles of writing discussed in the pages that follow—they will need greater discernment to review and revise texts generated by machines. Chapter 8, "Technical Writing," includes a few brief thoughts on writing with AI.

WHAT HASN'T CHANGED?

Despite some structural changes and new content, the fundamental premise of the book—to be effective, naval writing (and professional writing in general) must carefully target the intended audience and situation in content, structure, and style—remains central to all of the guidance we provide.

The Naval Institute Guide to Naval Writing still heavily incorporates practical advice and experience from the fleet. I have aimed to maintain Bob Shenk's informal, conversational tone. With the exception of chapter 1, the fourth edition continues to use a chapter structure based on specific document types or categories of naval writing.

Many of the document examples from earlier editions are here updated or adapted. Some examples are newer, but I have also retained many of Shenk's original, excellent examples to demonstrate the timelessness of effective writing principles and to preserve some connection with the Navy of the late twentieth century. I hope you will find this edition a useful tool in communicating, leading, and getting the work done.

Before getting started, I'd like to acknowledge and thank a number of people for their help with this edition. LT Morgan Cope was a tremendous help with reviewing content of several important chapters, and she put me in touch with several additional contacts. James Nierle and Lisa Wellman were very helpful with the "Awards and Commendations" chapter. MC1 Mark Faram provided useful guidance with the news and social media chapter. CAPT Curt Phillips (USNA 1990) offered insight into operational writing for the Navy. CAPT Donnie Kennedy and CAPT J. P. Garstka provided invaluable help updating the examples and advice on evaluations and fitness reports (fitreps). In addition, I'm grateful to Jim Tanner, CAPT Mike Schwerin, Paul Wilder, CDR Mike Flynn, LCDR Mike Major, and

MAJ Jeff Ludwig. Thanks also to Min Suh Lee, who helped with reformatting my initial files from the third edition.

The greatest debt of thanks goes to Bob Shenk, who passed away while I was working on this edition. We never met in person, but we exchanged emails and a phone call early in the process. He generously gave advice for ways to update the *Guide to Naval Writing*. The tremendous work he invested in creating the first three editions remains the foundation of this fourth edition. His wit and conversational, clear style remain present.

Finally, thank you to my wife, Sonmin, for the encouragement I need to do anything, and to my children. In particular, thanks to Elliott Crane (an ROTC midshipmen at the time, now an ensign), who provided early input on needed content revisions to several chapters, and to Olivia Crane, who proofread several new sections thoughtfully, catching many errors, offering insightful suggestions, and employing a careful eye for formatting.

C. E. CRANE

Introduction

You might think that in our highly digital, highly visual tech culture, writing skills would take a back seat to things like coding, graphic design, or photo editing. In fact, the opposite is true. Our increasingly digital culture gives us more ways of writing—and need *for* effective writing—than ever. Yet true skill in the art of clearly articulating an idea for the intended readers and purpose seems all the more scarce. In the multifaceted, multimodal professional world of today's Sailors and Marines, knowing how to put thoughts, plans, questions, intentions, reports, updates, instructions, news, requests, policies, and commands into the written word with clarity remains as vital to the Navy's and Marine Corps' missions as it has ever been.

WHO DOES THE WRITING IN THE NAVY?

"Most of the paperwork falls on the shoulders of those who can write—which usually means those individuals [who] have lots of face time with the XO and CO."
—LDO, ADMINISTRATIVE OFFICER ON A CARRIER

It is not surprising, perhaps, that the writing done by senior officers carries weight. But junior personnel, both officers and enlisted, undertake a major portion of the writing in the sea services. The duties of several Navy and Marine officers, petty officers, or sergeants; a Marine major; and civil servants interviewed for current and prior editions of this book stand out. Their responsibilities demonstrate how important writing can be at almost any level. Officers and senior enlisted must daily read, respond to, forward, or otherwise deal effectively with dozens of official emails from every conceivable naval addressee. Many of the specific genres of documents mentioned here and in the chapters that follow are now sent via email, the understanding and use of which requires considerable craft of its own.

For example, a lieutenant, junior grade was made administrative officer of a destroyer late in his first shipboard tour because he was such a good writer. He ended up writing all of the award nominations, many evaluations, frequent press releases, and much of the command correspondence on board ship. He said he gained invaluable perspective from working directly for the commanding officer (CO) and executive officer (XO).

A chief boatswain's mate was assigned as the first lieutenant on a small ship, writing many instructions and reports, recording some counseling sessions, and even putting together the report of a *Manual of the Judge Advocate General* (JAG-MAN) investigation (see chapter 9) in addition to evaluations and award narratives.

Similarly, a disbursing officer on an aircraft carrier wrote a great deal in addition to the money counting and number crunching the Supply Corps trained him to do. Like the chief, this junior officer (JO) was assigned a JAGMAN investigation shortly after reporting on board. He also wrote letters of commendation, award justifications and citations, and letters explaining complicated tax exclusions and combat-zone pay. Many of these documents went to his CO to sign.

In a final example, a Marine administrative sergeant happened to work alongside officers who drafted replies to service members who had written the president. When the officers suddenly got overloaded, the sergeant was asked to contribute. Soon he was researching and drafting eight to ten letters a week, many for the signature of the president of the United States. This experience later served him well in the adjutant's office of a Marine division, where he supervised Marines writing responses to congressional inquiries.

What Is Naval Writing?

Virtually all Navy and Marine Corps professionals need to learn the principles of good writing early in their careers.

Nearly every role at every rank and grade requires you to write emails. Composing and organizing these messages, as well as memos and letters, properly is vital to getting daily work done. Point papers and briefing folders circulating internally on staffs, both electronically and on paper, articulate policies and procedures, while budget justifications and technical reports help build and maintain the services. Staff reports and formatted messages feed and equip fleet units, while reports sent by operational units back to a rear staff inform commanders of the situation in the field. Without the messages, after-action reports, lessons learned, logistics requests, and many other documents that originate from ships and Marine operational units, shore staffs would have little information with which to form impressions, solve problems, and make changes to guide the Navy and Marine Corps.

When junior enlisted become senior noncommissioned officers, they will have to draft enlisted evaluations and letters of commendation for their people and will contribute to letters and reports. JOs will do the same, and sooner or later almost all of them will join staffs on which writing becomes a major personal responsibility.

Personnel evaluations ensure that the services promote the best individuals and help place the right people in key positions, while award justifications, letters of commendation, and award citations recognize individuals for their achievements. At the very least, almost every officer has occasion to write recommendations for enlisted men and women leaving the service. If you don't know how to address civilians about the talents of your subordinates, your letter probably won't help your people very much.

Social media posts for a command and news features in base papers help keep up morale and spread the word about the good jobs people are doing. News releases, letters responding to congressional inquiries, and formal position papers present information and naval perspectives to the general public and to public officials.

No official guidance tells you how to communicate with a previous CO or other official on missing evaluations or fitness reports or how to request the removal of evaluations from your record. Furthermore, *whether* to write to the selection board may be argued about at happy hour, but *how* to do so seldom appears in print as it does in this guide.

With confidence in writing and the basic principles and examples in this book, you can adapt to a variety of writing situations. You cannot prepare for anywhere near all the specific documents you'll have to compose in a naval career, nor can this text offer instruction on every possible situation or document. But you can master the principles of good writing, practice them in the situations we discuss, and with these same principles to guide you, adapt to other circumstances by using your head.

1

Effective Writing Fundamentals

"JOs are pretty sharp. They are weakest in writing skills, especially newly commissioned officers. Specifically, they still try to write like 'college students,' with flowing paragraphs and big words that impress English professors but not evaluation/correspondence receivers."

—COMMANDING OFFICER, NAVY

EXECUTIVE SUMMARY

Writing effectively in the professional world of the Navy and Marine Corps (as in the civilian world) requires you to account for some elements not often covered in one's many years of writing papers in school. Navy and Marine Corps writers should tailor documents for each specific audience, purpose, and context. As one of those writers, you also need to consider your relationship with a document's readers during composition. Understanding writing as a process with many steps rather than simply as a product can help you plot a course through that process. "Writing" involves more than composing the sentences that end up in the document. It requires determining the best content and organization for the ideas and figuring out a central focal point—a "bottom line" as we often say. Drafting the text, revising the writing style, and final proofreading and editing are also distinct steps, which may even include changing elements of content and structure. Drawing on external sources of expertise—military and civilian, in print and online—can also help make your writing successful.

CHARTS AND INSTRUMENTS: NAVIGATING THE WRITING PROCESS

In keeping with the nautical theme of this book, and at the risk of overreaching in the art of metaphor, I offer some principles of writing in a maritime analogy. In traditional navigation at sea or on land, you need fixed points of reference from which to plot your positions and direction (although GPS does most of this work for us now). Similarly, writing effectively in the Navy and Marine Corps does not occur detached from external factors. It requires an awareness of many aspects of the situation. Furthermore, writing doesn't begin when you start composing sentences and paragraphs. It is a process that starts as soon as you become aware of a need or a problem that requires some form of communication as part of the solution. The writing process begins when you realize this need and start thinking about the issue, the reason for writing a document.

TAKING BEARINGS: A WRITER'S TRIANGULATION

During the process, before you make decisions about the order of presentation or about whether to use "determine" versus "figure out" or "continuous" versus "continual," take bearings on the writing situation itself. The classic writer's (and speaker's) triangle is depicted in figure 1.1.

The idea captured in this illustration is that you (the writer or speaker), the ideas and information you are putting into words, and the people who will read or listen are all distinct but closely connected elements in determining what you say and how you say it.

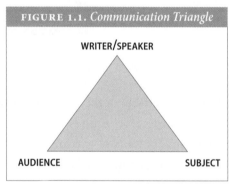

FIGURE 1.1. *Communication Triangle*

Understanding these elements for a given situation is essential to success in any real-world writing situation (as opposed to composing papers for school, where in most cases you are in training and are writing to show a teacher or professor what you learned). By knowing your audience, your subject, and yourself (including how others will see you in the writing situation), you can gauge your position with respect to any particular communication. Many writers are highly attuned to these elements when texting friends or posting on social media, but their written and oral communication at work often reveals a lack of understanding about that context.

CONSIDER YOUR AUDIENCE

The most important principle for effective writing is to adapt your content, structure, and writing style for your specific audience. In any communication you should account for the perspective, needs, situation, and background of the person or group you're addressing. For instance, knowing a captain's preferences toward granting leave requests or allowing exceptions to rules might affect the way you word an email asking for time off or special permission. Your readers' level

"I teach this course to help make you not look like an idiot in front of your CO."

—RETIRED COMMANDER, INSTRUCTOR OF PROFESSIONAL
WRITING, U.S. NAVAL ACADEMY

of technical knowledge and concerns about the issue should also factor into your decisions about what to say, how to structure your document, and how you put it.

Taking your audience into account can also help you adjust your style. How technical or specialized the wording can be depends on whether your readers will understand such terms. Unless they will all be aviators, for example, avoid terms like "painting bogies." Even if writing to aviators, you shouldn't use such informal jargon in official correspondence that leaves your command. You don't know what secondary audience will read it and view such terms as too informal or technical to be used outside the briefing room. Nor do you know whether an informal email (with a joke or gossip) will be forwarded at the touch of a key to third parties you never considered addressing.

Rank and position play major roles in professional relationships and communications. In the military the position and role of your reader also factor into the writing situation: is she or he higher up in the organization, at your same level but in a different branch, or in a different organization altogether? For example, whether you're writing to someone senior or junior to you should make a difference in the tone of a letter or memo.

You should also consider the role and rank of the boss of the people you're writing to independent of your immediate recipients' relationship to you. Knowing the standard policy or positions on important issues of the unit they work in can be vital, too. Your attention to such things can affect, for instance, whether your proposal or memo receives standard routing, gets put on the back burner, or merits immediate command attention (for better or for worse).

In addition, writing down the chain of command in response to a personal request requires special care. It's easy to fall into the habit of patronizing a service member, speaking in stilted, bureaucratic terms or saying things like "perhaps you don't realize that the regulations require. . . ." YNCM Charles E. Miller Jr. (who taught the Action Officer Course at the Bureau of Naval Personnel [BUPERS] several years ago) regularly recommended that with such an audience you should "write like you talk—explain things at the reader's level. Don't sound like a Navy Regulation."

Writing to nonmilitary audiences requires even more adjustment and some simplification. As a naval official writing to a civilian, you'll usually have to explain more background than you do to fellow service members; you may need to do without all those Navy-specific or bureaucratic terms we all use so widely in-house. In short, thinking about the perspective and needs of those you are writing to is the very first principle of good writing awareness.

UNDERSTAND YOUR SUBJECT

It should go without saying that, at least by the time you finish writing your document, you should know your material, your content, and your argument inside and out. Lapses in logic or in documentation, failures to explain or exemplify, or omissions of crucial details can sink your message, request, or recommendation before you get out of harbor. Because of space limitations, very little of your knowledge may show up in the body of a briefing memo. But in tabs, appendices, links, or material held in reserve, you should be able to support any position and answer any question from your audience. Make use of examples, statistics, pertinent testimony, comparison and contrast, and definition and analysis—all the equipment of standard argument and explanation.

KNOW YOURSELF

You don't need the Oracle of Delphi to advise you of the importance of this third point of triangulation. Always remember who you are in relation to your audience and the role or persona you want to project when composing your letter, message, or speech. Sometimes you may write as the division officer, other times as a watchstander or in the role of a collateral duty you hold. Even emails will vary, sometimes being in your own voice simply as a fellow petty officer, chief, or officer in the command; at other times coming from you as the ship's Combined Federal Campaign coordinator or perhaps in the coveted role of command urinalysis officer. As with audience, your choice of content, structure, style, and tone should be guided by who you are in relation to your readers (or listeners) and your subject.

Some service members forget this principle when they compose letters to selection boards, criticizing the service or their superiors to explain away some low rankings. Such writers come across as malcontents to board members, which may hurt candidates far more than their low rankings. Be aware of the impression you make. If you are appearing before the CO or an admiral with a budget request, speak boldly about the needs of your division or command. On the other hand, as your ship's liaison to families discussing how the deployment schedule affects spouses, pay, and children, speak in a more gentle, compassionate tone. Do the same in writing.

Clearly, deference and respect are always good qualities in juniors writing or speaking to seniors. But you will sometimes be in a position to speak for the command, not just for yourself. Even if you're relatively junior, when representing your command at a conference, for example, you can be forceful in your expressions of opinion. E-7s or E-8s from a staff may speak decisively and with strong credibility even among senior officers, because they know the admiral will back them up. Naturally, seniors are expected to be confident—falling short can undermine your credibility. Act, speak, and *write* so as to be believed and respected in all contexts. You may not care about that respect, but for the sake of the chain of command and the mission, you must cultivate it professionally.

WRITE FOR YOUR PURPOSE

Few people in the workplace write memos, directives, or email simply because they are fun exercises or because they have nothing else to do. We write to meet a need, to solve a problem. Even a memorandum for the record (discussed in chapter 2) has an audience of your future self and anyone with a need to read what you have documented. In your writing, are you aiming to persuade your reader or simply to provide information? What do you expect or hope will happen after your document is read? Does it need to create a change of behavior or a new way of thinking? The answer to those questions should shape what you say, what details you emphasize over others, your structure, and the tone and style your comments.

ACCOUNT FOR CONTEXT

"Nobody reads at a desk in a quiet room."

—RETIRED COMMANDER

Anticipating the way people typically read a given type of document can also be valuable. Naval directives are usually scanned, not read word for word, so headings can help alert readers to key information. On the other hand, extremely technical material requires great patience on the part of a reader, so the writer of a manual must be careful to explain clearly and meticulously and provide many aids—summaries, headings, visuals, section reviews, and so on. The large number of emails handled daily by naval offices everywhere suggests the critical importance of accurate and descriptive subject lines.

Second, you must make adjustments for the subject matter at hand. Those who live or die by whether what they write is actually read—recruiters, for example—know this principle very well. "As a recruiter I learned that as soon as my recruiting document went beyond a paragraph or a paragraph and a half, it was thrown away," observed a Navy chief. "So, I had to get a reader's attention in that opening paragraph." Depending on the subject, a writer doesn't always need to be so immediate with the message. Most of us will study an explanation of changes in our pay and benefits with more care than we would give to the email announcement about the family fun fair on base some Saturday afternoon.

Reading the Wind: Situations and Contexts for the Naval Writer

WRITING FOR THE BOSS

"The personal correspondence that you will draft for the admiral presents a new and interesting problem. You must write in your admiral's style and vocabulary. You'll have to learn what tone and level your admiral would use in different situations and with different people. You must be able to convey a portion of the admiral's personality."

—*NAVAL FLAG WRITERS HANDBOOK*

You will often find that you don't write for your own signature but for someone else's. In many an email, letter, directive, or report, the drafter disappears, and the only name that appears (or is assumed) is that of the CO. This principle is even truer on major staffs, whose action officers write for very senior officers.

A chief yeoman commented that even after special training to write for his two-star admiral, he needed six months to really learn to write like the admiral before his work began progressing smoothly. Whatever time it takes, sooner or later not only specialists like flag writers and staff action officers but also almost all naval professionals must learn to write for their superiors. A lieutenant commander, the XO of a helicopter squadron, said he often encountered difficulty in writing for "the command." Having learned through some practice, he made this suggestion: "Keep the facts in, and leave the adverbs out." He also advised giving the skipper more than was needed to allow her to decide what was necessary, simply deleting any extra parts. Not providing enough information makes the CO come back to you to complete the paperwork; if all she has to do is chop something out, she can give the document to the yeoman immediately, thereby cutting out an extra step. This process may vary greatly from one CO to another, however.

The situation of writing for the boss (or bosses) can be extremely complex. After working in a high-level bureau in the Office of the Chief of Naval

Operations (OPNAV), a commander reported: "I became somewhat frustrated in my previous job because the deputy director and director had radically different writing styles from myself and each other. The result was an awful lot of wordsmithing."

She then went on: "I think that every new tour puts you in the position of writing for a new senior who has his own writing style—good or bad. You have to learn the basics of expressing yourself clearly and succinctly in writing, but you have to learn to be flexible and not to take the rewriting by your seniors too personally. The bottom line is to find the best way to communicate your points effectively so that you get the support or approval you're seeking."

Indeed, seniors should assign juniors to write for them not only to reduce the workload but also to help juniors learn. This practice helps improve a junior's writing style and teaches a commander's perspective. For example, drafting night orders for a destroyer's CO while the ship is underway can become an exercise in thinking as the commander has to think, in seeing through the skipper's eyes, as it were. If you see how the CO modifies what you've written, then you've shared a commander's general outlook and operational perspective.

See chapter 3 for more detailed guidance on writing for others on a staff.

WRITING AS THE BOSS

Writing *to* the boss can be a challenge, but writing *as* the boss is no easier, whether you are a CO, XO, a division officer, chief, or other supervisory role—or as a junior writing on behalf of and in the voice of the boss. In this context your readers are those who are under your authority, so you should keep the following principles in mind.

Your troops might be a little afraid of making mistakes, of displeasing you, so they might try to read everything you write very literally. Take care to make clear what you consider "rules," or specific expectations, and what you consider general guidance, or "commander's intent."

Your tone and voice may come across as much stronger or harsher than you intend in writing simply because of your seniority and positional authority. For example, if you *feel* a little irritated when you are sending an email or putting out new guidance, you will likely *sound* quite angry to your troops.

Make clear in your writing both the general idea and the supporting details, the specifics. This principle is important in all written communication.

Don't expect your troops to read between the lines. Articulate (or have your staff writers articulate) your expectations explicitly. Spell things out clearly for the least knowledgeable, least intuitive person in your audience.

Charting Your Course: Content and Structure

"Don't fine-tune all you write, but only the most important things. Get the work out. Adapt existing documents to your purposes—don't reinvent the wheel."
—LIEUTENANT COMMANDER

Awareness of the basic communication triangle (Figure 1.1)—the three situational fixes from which to plot your course—and of the vital importance of accounting for your purpose and your context provide the foundation for any writing or speaking situation. Yet the writing process comprises many steps: researching, brainstorming, drafting, revising (sometimes multiple times), and editing. Some may take only a few moments, while others may take days or weeks. Understanding writing as a process and not simply as the end product can help you better navigate that process and produce an end product that better achieves your intentions.

RESEARCH YOUR CONTENT

Take account of all the notes, documents, directives, messages, and other written information you already have on hand. From that, consider what more you require, then do the necessary research. Unlike your academic research in school, however, research in the professional setting usually involves reading background documents and talking with people involved.

When writing messages on board ship or ashore, the standard means of research is to search through previous messages and other shipboard files. When writing naval letters, one often looks to directives, office files, regulations, manuals, policy letters, and operating instructions. Writing a performance evaluation, on the other hand, primarily requires knowledge of a service member's performance. Similarly, with award nominations, the research required is to get the facts about what an individual has done. JAGMAN investigations can call for a variety of kinds of research, from conducting interviews to looking into regulations, evaluating procedures and logs, and so forth.

You may also need to conduct library or online academic research in naval writing—perhaps when writing a professional article or drafting a speech. Guided by a standard tool such as the Air University Library Index to Military Periodicals (AULIMP, available online), you can look into past articles to see what has already been said about your subject.

Your situation, subject, and writing context govern the kind of research you must do. Be sure to talk to experts at your command or others nearby about new writing tasks. For example, see the XO or a senior yeoman about writing letters and directives, consult a local staff judge advocate on JAGMAN investigations, or see members of the local staff "secretariat" or civilian consultants when you work on major staffs. Don't forget to consult past files of reports of the same kind you have to do and the people who put those reports together, if they are still on station or reachable by email inquiry or phone. These sources can offer invaluable suggestions on where to start and how to proceed.

Of course, there are other brainstorming, freewriting, or first-drafting techniques, all of them methods of "rhetorical invention," finding the important things to say and then beginning to say them on paper. These launching techniques have been studied in depth by modern communicators; if you've been taught a particular method in a composition class, go to it. Use whatever launching technique works best for you.

DRAFT AN OUTLINE

Outlining is one of the classic ways to begin. Once convinced you have enough material (admittedly, sometimes hard to tell), prepare to compose. At this point, unless you have only a short memo to draft, pause long enough to create a rough outline. In our nautical analogy, creating a good outline is like plotting the course on a chart or on the computer. A plan for the structure of a document will help you step back and look at whether the pieces of your discussion, explanation, request, or argument will really get you to your destination—the overall purpose of the document. Don't make it so formal that you focus more on outlining than on writing, but use it to set your sights.

By going through your notes and listing specifics in an outline, you're checking the chart and casting off lines. Once you have an outline prepared, you can usually begin drafting sentences. Creating an informal outline can feel repetitive; you may feel you are wasting time. But the initial thinking through of structure will pay dividends later in terms of clarity and cohesion of your end product and saving time during any revisions.

OPEN WITH A SUMMARY

Writers use summaries to preface technical reports, research reports, correspondence packages, and many staff documents—but we could use summaries more widely still. Several common methods of summarizing are highly useful for naval writers. Briefing memos, executive summaries, and abstracts find wide use at commands and staffs, as they help busy senior officials quickly get the gist of any particular document. And letters of transmittal are helpful in pointing out key information in the complex documents or the thick correspondence packages that they accompany.

Also useful are the standard ways of organizing the material that follows the summary. For example, consider the inverted pyramid used by news writers to organize their articles following the news lead (discussed in chapter 10). This technique accustoms journalists to state the vital facts first, then to add other information in descending order of importance, a habit that can be useful in naval documents as diverse as operational reports (OPREPs), performance evaluations, briefing memos, and longer documents sent by email.

GETTING UNDERWAY: FROM THOUGHTS TO WORDS

Preparing an initial draft of your main content, however rough, is a vital next step. Don't let anything stop you from getting the bulk of your argument down on paper quickly. Focus on major ideas and points, fleshing out your outline. Avoid getting sidetracked by looking up specific dates, figures, names, or any technical details. Getting these right is important, of course, but at this point keep your forward momentum in writing. To put it in nautical terms, keep your weigh on.

So don't consider grammar, spelling, punctuation, or style while writing the basic draft. You may know that you tend to write in an overly technical or bureaucratic style that can hinder clarity and directness and add needless bulk—don't worry about that problem now. You can go back and fix such weaknesses of style as well as grammatical errors later.

Even if your structure still needs work, keep writing the sections you can write about. Ideas can come and go very quickly and may not return. Get them down in draft form, waiting to polish and revise them later. Capture the raw material and leave the finished product until later.

Once you've got a mostly complete draft on paper (or on the screen), take a bit more time to go back and clean it up. Fill in the names and figures (always double check the figures), flesh out passages you left sketchy, correct the errors you notice, add necessary transitions, and look up additional information as needed. Make that first draft as complete a draft as possible even though it will still need revision.

ADD VISUALS, IF APPROPRIATE

Charts, graphs, and drawings may help explain your topic or convince your audience. Such aids are normally most appropriate in oral briefings, technical reports, and professional articles rather than in correspondence. But they may be added as appendices to correspondence packages and occasionally for instructions. Make sure they are simple and clear in presentation—cluttered visuals only confuse and slow the reader.

CHECK THE VISUAL IMPRESSION YOUR DOCUMENT MAKES

Make sure the font is suitable, the type is clean and dark, and the margins are reasonable—normally one inch on all sides. The sharp appearance of paperwork—whether actually printed out or sent by email from computer screen to computer screen—can make as much of an impression as a sharp uniform and be even more important.

LET IT REST A WHILE

One of the keys to a good revision is a rest—time for your composition to stew. A limited-duty officer working for a three-star in OPNAV remarked in frustration: "If they'd only let the paper ferment a half hour, anyway. But they're caught up in the rush, and as a result they don't get to the major topic until page three."

On a large writing project, let your first draft sit for a day or two, if possible, before you come back to it. If you simply don't have the time, at least try to wait thirty minutes before going further. You'll often be surprised how different your text will look with just a half-hour separation from it. Even short emails can often profit by a short period between drafting and sending.

TECHNIQUES FOR ORGANIZED WRITING

Start Out with Your Main Point

Except in rare circumstances in which you want to "talk" a reader into an idea or seek to soften the blow of unwelcome news, put your "Bottom Line Up Front" (BLUF). As the CO of a recommissioned battleship once commented: "Make your bottom line your top line; put your main point up front. Flag officers don't have time to read anything but the very key points." From there, put supporting points in descending order of importance.

To get the reader's attention from the start, make the subject line (in a document with a subject line, like a letter or an email) as detailed as possible in a few words. Write "Eliminating Restrictions in Camp Lejeune Training Areas" rather than just "Training Areas" or "Seabee Participation in Fiji Aid Program" rather than just "Aid for Fiji."

When you write a document, think about the sentence you would keep if you could keep only one. It should appear by the end of the first paragraph—ideally, even earlier. Some of the clearest document structures highlight the main point in a one-sentence paragraph at the very beginning. Put requests before justifications, answers before explanations, conclusions before discussions, summaries before details, and the general before the specific. Avoid mere chronology.

Occasionally, you might delay your main point to soften bad news, for example, or to introduce a controversial proposal. But in most cases, don't delay—plunge right in.

To end most letters, just stop. When writing to persuade rather than just to inform, you may want to end strongly with a forecast, appeal, or implication. When feelings are involved, you may want to exit gracefully with an expression of good will. When in doubt, offer your help or the name of a contact.

Use Topic Sentences

A paragraph may need a topic sentence, a generalization explained by the rest of the paragraph. Then again, it may not. In a short paragraph, a topic sentence may be unnecessary if a reader can follow the writer's thinking without one.

Be alert to the advantages of a topic sentence, as it helps shape masses of information. Without one, some paragraphs make readers shrug and say, "So what?"

Use Short Paragraphs

Long paragraphs swamp ideas. Cover one topic completely before starting another, letting a topic take up several paragraphs, if necessary. But keep paragraphs short, roughly four or five sentences. Call attention to lists of items or instructions by displaying them in subparagraphs.

Now and then, try a one-sentence paragraph to highlight important ideas.

INFORMATION DESIGN AND DOCUMENT NAVIGATION

A valuable tool for guiding your readers through your structure—the relationships of the parts to one another and to the whole—is the use of common visual formatting. White space, bullets, numbered lists, headings and subheadings, and other options for how the words look on the page or screen can greatly help your audience.

Headings

Headings can greatly help readers skim a text, find material, and simply comprehend. Consider using them on anything longer than a couple of paragraphs (including reports, staff documents, and directives). Use headings even in emails of more than a paragraph or two if you need to catch a reader's eye.

In addition, make them as interesting and pointed as possible to draw readers' attention. Make sure they're grammatically parallel—match a full sentence with a full sentence, a noun phrase with a noun phrase, and so on—within each level of heading. See the headings in this text as examples.

Bullets

A "bullet" is a piece of type used to introduce each element of a list or series; it is shaped like a bullet seen head-on: •. You can, of course, use other symbols available in your word processor for these. People also refer to phrases and truncated sentences as bullets with or without a symbol in front.

Bullets should be grammatically parallel in form and in sentence order, like this:

- Bullets help a reader skim.
- Bullets emphasize key ideas.
- Bullets often enliven a text.

If you use sentence fragments instead of complete sentences, head them with an introductory statement. Ensure that each such bullet

- completes the clause,
- contains the same basic structure, and
- expresses its content concisely.

Ensure that you follow any in-house formatting guidance, and don't use symbols that are so unusual that they draw attention to themselves. Remember to use strategic placement in any list of information. Normally, the first item in a series of bullets (and to a lesser extent the last) will attract the eye and be much more obvious than an item buried in the middle.

Other Visual Formatting

Additional tools for information design to draw attention visually to particular data are boxes, -pointers-, *italics*, vertical spacing, **bold print**, larger font, or ALL CAPS. Of course, there are stars, pointing hands, and many other devices available electronically. The degree to which you can freely use such visual formatting varies with the formality of the document you're writing and with the audience for whom you're writing.

Of course, don't overdo your use of typographic devices or you'll make your document more difficult to read (rather than less), lose the effectiveness of the device or formatting, and risk accusations of being "cute." Overuse of these features and changes in size of type also can make a page resemble an old-time billboard advertisement for a circus rather than a professional document.

Trimming Sails:
Revising and Proofreading for Plain Language, Style, Tone, and Grammar

"You certainly need to use a thesaurus—I highly advise it. But you have to have a
vocabulary to use one in the first place!"

—MASTER CHIEF PETTY OFFICER OF THE NAVY

PLAIN LANGUAGE

The Plain Writing Act of 2010 requires federal agencies, including the Department of Defense (DoD), to make writing that goes out to the public well organized (see the advice above), clear, and concise, using language appropriate to the subject matter and the intended audience. While much of your writing will be for internal audiences, the principles of plain writing (also called "plain language" and "plain English") are useful for all professional communication. And the move to write more in plain language is broader than the government; mastering this skill will serve you in many spheres.

Many people edit others' documents more often than they write their own. JOs and senior enlisted typically draft more than they edit, whereas their supervisors—department heads and XOs—often find themselves editing and proofreading much more than they write. All should know the basic principles of editing. The next several sections examine plain language at the sentence level, but the principles discussed earlier in this chapter, such as organizing a document clearly and formatting it visually for ease of navigation, are also part of writing in plain language.

Of course, not all of the tips that follow will apply to every document. On shorter pieces you can look for several problems at once. With written material several pages long or with an especially high-stakes or high-visibility document, we recommend reviewing the writing several times, looking for different problems with each reading. Don't try to revise for all of these elements of clear writing at once. Make one pass through your document looking at ways to improve sentence structure. Perhaps in another round of revision you focus only on changing passive to active voice where needed. A third round might focus on simply cutting out excess words.

We begin this discussion of plain language principles with some famous military sayings reworded into bureaucratic, unplain navalese. See if you can figure them out! See the end of this chapter for the original versions.

FIGURE 1.2. *Famous Military Sayings Converted to "Navalese"*

1. Argumentative contesting by originator has not yet commenced.

2. May the metallic underwater explosive devices be execrated. Let maximum velocity in a forward direction be achieved forthwith.

3. When the subordinates desire, permission is hereby authorized for the expeditious emission of ordnance.

4. Upon arrival in the theater of operations, an overview of the environment was conducted, and the conflict situation was subsequently resolved in my favor.

5. Expenditure of ammunition is to be withheld pending the detection of whiteness in the ocular organs.

Note: Number 4 was originally used as the epigraph to CAPT Carvel Slair, USN, "Effective Writing, Navy or Civilian," U.S. Naval Institute *Proceedings* (July 1968): 131. See Figure 1.4 for the original phrases.

SENTENCE STRUCTURE AND NATURAL STYLE

Craft Sentences for Emphasis

Here are four ways to write sentences that call attention to important ideas.

Subordinate Minor Ideas to Major Ones

For example, "*Although the change of command is delayed by one week*, the uniform inspection will still take place on April 25." Subordination clarifies any complex relationship between ideas.

Place Key Words Intentionally

You emphasize a word or idea when you place it at the beginning or end of a sentence. To mute or deemphasize an idea, put it in the middle of a sentence.

Use Parallel Structure for Parallel Concepts

Look for opportunities to arrange two or more equally important ideas as a pair or list. Parallelism saves words, clarifies ideas, and provides balance. In any series (list), the items in the list should be of the same grammatical type, such as verbs, nouns, phrases, or whole clauses (for example, the verbs "saves," "clarifies," and "provides" in the previous sentence and the use of a verb to start each part in this section).

Try Some Mini-Sentences

An occasional sentence of six words or fewer slows readers down, creates rhythm, and emphasizes ideas.

Speak on Paper

Make your writing as formal or informal as the situation requires, but do so with language you might use in speaking. Because readers hear writing, the most readable writing sounds like one person talking to another. Begin by imagining that your reader is sitting across from your desk. Then, write with personal pronouns, everyday words, and short sentences—the best of speaking.

Use Personal Pronouns

Though you needn't go out of your way to use personal pronouns, you shouldn't go out of your way to avoid them in most documents. Avoiding natural references to people sounds artificial. Speak of your activity, command, or office with "we," "us," or "our" in most correspondence. These words are more exact and natural sounding than the vague "it." Use "you," stated or implied, to refer to the reader. Use "me," "my," and "I" in emails, but deploy them less often in other documents. These personal pronouns are especially useful in correspondence signed by the CO to show special concern or warmth.

Rely on Everyday Words

Write to express, not to impress. The complexity of our work and the need for precision can require some highly specific, precise terms. But in many cases, a simpler or shorter word will do just as well as a "big" one. For example, deflate "utilize" to "use," "commence" to "start," and "promulgate" to "issue." Favor short, spoken transitions over long, bookish ones.

Avoid the needless complications of legalistic lingo; let a directive's number or a letter's signature carry the authority. Use "here's" instead of "herewith is" and "despite" instead of "notwithstanding."

Use Contractions Cautiously

Contractions link pronouns with verbs ("we're," "I'll," "you're") and make some verbs negative ("don't," "can't," "won't"). Since contractions are less formal, you should judge whether using them fits the context and audience expectations. Much day-to-day naval writing such as routine emails should be informal enough for contractions to fit naturally. If you are comfortable using them, your writing is likely to read easily, because you will be speaking on paper. Note, however, that some formal context and document types—a ship's instruction or a naval directive, among others—should not use them.

Keep Sentences Short

For variety, mix long sentences and short ones, but try to keep the average under twenty words. Though short sentences won't guarantee clarity, they are usually less confusing than long ones. Try the eye test: your sentences should average fewer than two typed lines. Or try the ear test: if you can't finish reading a sentence aloud in one breath, break it into two sentences.

Phrase Requests as Questions

A request gains emphasis and sounds like natural speech when it reads as a question and ends with a question mark. Consider the following examples:

> Request three additional personnel be temporarily assigned to the galley to support the food preparation for the air wing.

> Would you approve a temporary assignment of three additional personnel to the galley to support food preparation for the air-wing embarkation?

Do you hear how the second version, as a question, more naturally invites the reader into a kind of relationship or partnership with the writer? A request in the form of a question can also more effectively convey the military deference to the authority of the person you are making the request to.

CONCISE, CLEAR WRITING

Cut the Fat

Give your ideas no more words than they require to convey your meaning clearly and naturally. The longer you take to say things, the weaker you come across and the more you risk blurring important ideas. Suspect wordiness in everything you write. When you revise, tighten paragraphs to sentences, sentences to clauses, clauses to phrases, and phrases to words as much as possible. To be easy on your readers, you must be hard on yourself.

FIGURE 1.3. *A Memorandum Excerpt Revised into Plain Language*

Original Memorandum:

1. Your request for $750, as stated reference (a), is approved. Contact Mr. John Jeffries, X4442, to coordinate the appropriate procedures for utilization of funds from Account 158 of the Foundation Support Fund.

2. Liaison with the Foundation Support Committee Associate Budget Officer indicates that this item will be supported out of departmental funds in the future, and, therefore, is not considered a routine budget item for Foundation Support Funds.

Revised Memorandum:

1. You can have the $750. John Jeffries at X4442 will tell you how to get the money.

2. In the future, however, your department will have to pay for this, instead of the Foundation Support Fund.

Figure 1.3 shows a memorandum one naval office sent to another, along with a revision of that document by someone fed up with "budget-speak." Notice how problems of passive voice, long words, and officialese plague the original version. The writer should shorten it, make it more direct, and speak more naturally.

Be Concrete

Without generalizations and abstractions—lots of them—we would drown in detail. We sum up vast amounts of experience when we speak of "dedication," "programs," "hardware," and "lines of authority." But lazy writing overuses such vague terms. Often writers weaken them further by adding adjectives instead of detailed examples: for instance, "*immense* dedication," "*enhanced* programs," "*viable* hardware," and "*responsive* lines of authority." Don't use a general word if the context allows for a specific one; be as definite as the situation permits.

Avoid "It Is" and "There Is"

No two words hurt professional clarity more than "it is." They stretch sentences, delay getting to the real subject and other important words, hide meaning and responsibility, and encourage passive verbs.

Similar to "it is" openings, forms of "there is" make sentences and subordinate clauses start slowly. Instead of opening with, "There are two possible courses that the ship could take during the exercise," write, "The ship could take two possible courses during the exercise."

Prune Wordy Phrases

Wordy expressions don't add meaningful weight to writing; they clutter it by adding bulk and getting in the way of the words that carry the meaning. "In order to" and "in accordance with," for example, are minor ideas that rarely merit three words. Instead of writing "It is interesting to note that," just start the sentence with whatever would come after "that."

Free Imprisoned Verbs

The most important word in a sentence is the verb, the action word that shows what the subject is doing. Weak writing uses weak, nonspecific verbs—often verbs of being like "is," "was," or "were," which usually require additional words to add meaning. When you use a general verb such as "is" or "make," check to see if you can turn a nearby word into a verb. The noun forms of verbs can bog down your sentence. Most words ending in "-ion" and "-ment" are verbs turned into nouns (technically called "nominalizations"). This practice is part of normal English grammar and is often useful. But writers tend to overuse these nouns without real-izing it. Whenever the context permits, change these words to their verb forms. For example, instead of "perform development" just say "develop."

Additionally, instead of saying "is applicable to," say "applies to"; instead of "make use of," simply say "use"; and replace "do an inspection" with "inspect."

Eliminate Unneeded Doublings

Pairs like "advise and assist," "interest and concern," or "thanks and gratitude" may seem to capture fine distinctions and appear more precise to you as the writer, but most readers will not appreciate this and will just be bogged down with the repetition. And most of your documents do not need such a fine level of precision—especially not at the cost of including these additional words. Repeating a general idea won't truly make it any more precise.

Avoid Long Noun Strings

These are phrases made of several nouns in a row in which each becomes an adjec-tive for the one next to it. Though grammatically permissible, these phrases are extremely dense and are not generally reader friendly. Sometimes called "hut-2-3-4 phrases," they exemplify being *too* concise. Readers can't tell how the parts fit together or where they all will end.

We must live with some established hut-2-3-4 phrases such as "standard subject identification codes" for subject codes, but you can keep them out of whatever you originate by adding some words or rewriting entirely. For example, you could revise "security inspection review procedures" to read "procedures for reviewing security inspections."

Avoid Excessive Abbreviating

Extensive abbreviating is less efficient than it appears. Use abbreviations and acro-nyms no more than you must with insiders, and avoid them entirely with outsiders. Spell out an unfamiliar abbreviation the first time it appears. If an abbreviation would appear only twice or infrequently, spell out the term every time and avoid the abbreviation entirely. Intersperse synonyms like "the command" for the more cumbersome abbreviation COMNAVSURFPAC, for example. Put clarity before economy.

Listen to Your Tone

Tone—a writer's attitude toward the subject or readers—causes few problems in routine letters. You may pay special attention to tone, however, when the matter is delicate. The more sensitive the reader or issue, the more careful you must be to promote good will. Tactlessness in writing suggests clumsiness in general. When feelings are involved, one misused word can make an enemy. To avoid tactlessness, use positive language.

Use Passive Voice Sparingly—and Only on Purpose

The need to write in a more active voice deserves some extended discussion here. You may have had a teacher or a supervisor tell you never to use the passive voice. While doing so breaks no grammar rules, the passive voice is less specific and makes new information more difficult to absorb for the reader. Therefore, you should minimize using the passive voice, saving it only for those situations where it really helps emphasize what you need for your purpose.

Quick Review of Passive Voice

A verb in the passive voice uses any form of "to be" plus the past participle of a main verb:

"is," "am," "are," "was," "were," "be," "being," or "been"

PLUS

a main verb form usually ending in "-en" or "-ed"
(the "past participle" form of the verb)

For example:
Passive: The hatch *was secured* by Seaman Patel; *or,* The hatch *was secured.*
Active: Seaman Patel *secured* the hatch.

Passive: Lateral transfer requests will *be reviewed* by the board in March.
Active: The board will *review* lateral transfer requests in March.

Notice how the passive sentences emphasize the hatch and the transfer requests over Seaman Patel or the board.

Passive sentences are usually easier to write, and many writers default to the passive voice without realizing it. Often, passive sentences don't show who or what has done the action. If a doer appears at all, it follows the verb. But most passives in naval writing only imply the doer, which is sometimes a severe problem when the context doesn't make the doer clear.

Revising from Passive to Active Voice

When you see a passive sentence and decide to revise it to active voice, consider these standard ways of doing so. Not all of these will work for every sentence; the approach you choose will depend on the verb itself and your desired tone and style.

Put a doer before the verb:
>   Passive: The part must have *been broken* by the handlers.
>   Active: The handlers must have *broken* the part.

Drop part of the verb:
>   Passive: The results *are listed* in enclosure (2).
>   Active: The results *are* in enclosure (2).

Change the verb to something that means the same thing but is not passive:
>   Passive: Letter formats *are shown* in the *Correspondence Manual*.
>   Active: Letter formats *appear* in the *Correspondence Manual*.

Use the Passive Voice only When Necessary

Occasionally, passive voice is exactly what you need. Three general situations often call for passive voice over active:

- When the doer is unknown
- When emphasizing the doer is unimportant or the doer is obvious
- When you intentionally want to omit the doer

For example,

Presidents *are elected* every four years.	(doer obvious)
The hatch *was left* unsecured.	(doer unknown or we want to avoid blame)
The part *was shipped* on 1 June.	(doer perhaps unimportant to state)
Christmas *has been scheduled* as a work day.	(doer better left unsaid)

When in doubt, use active voice, even though the doer may seem obvious or unimportant. You will write livelier sentences (not "livelier sentences *will be written* by you").

GENDER IN WRITING: QUICK POINTERS

Naval directives and consideration for your audience suggest that, where possible, we use pronouns and titles that do not specify sex. Gender-free language in our current climate can help create a tone and foster a culture of professional respect for individuals that the Navy and Marine Corps seek to cultivate. But not every naval term with "man" is required to be modified. Overly awkward revisions of terms such as "seaperson" instead of "seaman" can backfire and draw undue attention to the effort not to offend.

Legal issues further complicate this subject. For instance, a Joint Chiefs of Staff directive stipulates that titles established in directives or law, such as "airman" or "Chairman, Joint Chiefs of Staff," should be used as is (though the directive counsels to select gender-neutral titles when establishing new positions). On the other hand, while laws originally limited all combat roles to men, those laws have now changed, and the exclusive use of male pronouns in reference to Navy combat personnel is no longer appropriate.

One other important factor is tradition. Tradition reaches deep in the naval services, and naval authorities wisely foster tradition, knowing the great part it plays in human communities. When women first entered the Naval Academy, the superintendent decided that the term "midshipman" would not be changed but that the proper terms to use would be "male midshipman" and "female midshipman." Kindred decisions have been made throughout the services. At last report, "helmsman" and "man overboard" remain standard terms even on ships with women permanently assigned, although those same ships now warn of "personnel working aloft." Clearly, universal rules are hard to come by.

In the face of such complexities, keep your wits about you. Realize that certain choices of pronouns or titles may offend, thereby creating obstacles to effective communication. Remember also that the specific way each offends may differ, depending on the situation and the audience.

Pronouns

Follow the advice outlined below (based on the *Correspondence Manual*, among other sources) when you need to revise single-gender references to gender-neutral ones.

Write directives as if talking to a group of readers or to one typical reader. Use "you," either stated or implied. Instead of "The young officer must take his training seriously," write "Take your training seriously" or just "Take training seriously."

Choose plural pronouns if possible, such as "they," "their," or "them." Replace "A chief can submit her request to her division officer" with "Chiefs can submit their requests to division officers."

Note, however, that the use of "their" with a singular reference noun, as in "Anyone can take their laundry off base for cleaning," is technically incorrect (pronouns like "anyone," "everyone," and "someone" are actually singular). Usage is changing, however, and many of the most recent guidebooks condone this practice, especially because "they" and "their" are also gender neutral. But don't be surprised if your XO, CO, or administration chief (who may have been taught differently) objects to this. In formal writing you will be safer to say, "All Sailors can take their laundry off base for cleaning."

Reword sentences to eliminate pronouns. In place of "The private should return to his barracks," say "The private should return to barracks." Instead of "None of the Marines was proud of his performance in the exercise," say "None of the Marines was proud of the exercise" (be alert, though, to subtle shifts of meaning, as seen in this revision).

Substitute articles for possessive pronouns. As an alternative to "Every petty officer must be assigned *her* watch station by Friday," write, "Every petty officer must be assigned *a* watch station by Friday."

You can use terms referring to a particular gender in reference to a particular person if you know the person's preference. It is often wise to use job titles that include both sexes, such as "service member" instead of "serviceman" or "chair" rather than "chairman."

You should also check with current Navy policy on pronoun usage, which may continue to evolve after this advice goes to print.

As a final note, just as with other revisions for style, don't allow concern over the pronoun number or gender issue to interfere with your rough drafts—wait until the end of the process to adjust all such references.

PROOFREADING AND FINAL EDITING

"Spell-checking also has its pitfalls, most of which center on homophones [sound-alikes] such as they're/there/their, your/you're, site/cite/sight, to/too/two, and the like.

. . . All documents, no matter how short they may be, should be computer spell-checked by the writers, of course. But writers should also not allow themselves to be lulled into a false sense of security, since spell-checkers are not perfect. They will not notify the writer of the errors in a sentence such as 'Their maybe two many problems to site before the counsel at it's meeting.' Six errors lurk among those words!"

—WILLIAM K. RILEY, EDITOR, ARMED FORCES STAFF COLLEGE

Always Proof Your Work

Ensure your document is technically correct in every way or it will likely be sent back or delayed. Proof all aspects of the document. As the "Guide to OPNAV Writing" once pointed out, "A so-so paper that is technically correct will frequently get signed off, but a paper that needs to be returned for a typo invites a closer look and additional corrections of every kind." Use your spellchecker and have others proof your work as well. Even if it's a rush project and you're hustling to get it all done overnight, don't forget to proofread.

Grammar and Punctuation

Do not overlook this important step in finalizing any professional communication, whether an email, a performance evaluation, or a social media post. Ensuring your grammar and punctuation are correct is like putting on the right uniform or having your insignia on correctly: it is a fundamental professional expectation. Of course, people sometimes make mistakes (just as they sometimes have uniform items out of place at inspections), but you should avoid the temptation to think, "they will know what I mean," and thus not give attention to proper proofreading for these basics.

Of course, a complete overview of the basic mechanics of standard American English is beyond the scope of this book. Your word processor can help; you should use the grammar-check and spell-check tools before submitting an official document. But remember that these tools will not always recognize what you are trying to say, and they may not catch misspellings if you have substituted one word for another (such as "statues" for "statutes").

See the *U.S. Navy Style Guide* (Appendix B) for specific Navy guidance, and check the *GPO Style Manual* or some of the other resources listed in Appendix A for more comprehensive guidance.

Prescribed Formatting

Even if the Navy or Marine Corps has no particular format for a certain document, your own command may have. Most letters, correspondence folders, and formal reports require a standard format, especially at major commands. (If a local format differs from any format outlined in this book, follow the local practice.) The quickest way to get a piece of correspondence on a ship or on a staff sent back to you is to make simple technical errors.

Proofreading Tips

A few specific techniques can help make your final proofreading more effective—and potentially less time intensive:

- Proofread in hard copy. Spelling and other typos tend to be more noticeable on paper copies than on computer screens. Moreover, some features of a programmed document do not appear on the screen but only become obvious when printed.

- Read the document in reverse, starting with the last sentence, then backing up one sentence at a time. This method forces you to look at each sentence in isolation and can help catch basic errors you might overlook when reading quickly for overall meaning.

- Read your document aloud. Doing so forces you to slow down and to hear as well as see what you have written. You will likely catch more errors this way, and you will also notice any of the stylistic weaknesses discussed above, such as overly long sentences or wordy, confusing phrases.

- Ask a colleague to review your piece to catch things you may have missed.

Anchors and Moorings: External Resources for Writing

In your professional writing, make use of any informal guidance that is available, either personal help from experts or written guidance such as those mentioned throughout this book. You can learn a great deal from such informal material. For example, the bullet format that we have become so accustomed to now was almost entirely an unofficial development. We learned to write that way from seeing and talking about others' innovations. Used judiciously, unofficial guidance—word of mouth from selection boards, good examples of a point paper you've seen, semiformal writing guides put out by major staffs—can give a writer invaluable help.

Can you sometimes plagiarize in the Navy or Marine Corps? We don't usually call it plagiarism, but to an extent, yes, you can; some written material really constitutes what we call "boilerplate"—material usable in a variety of contexts with little variation. For example, award citations typically begin and end with standard phraseology, and an award justification (the write up that makes a persuasive case for the person to receive the award; see chapter 6) can be built partly out of bullets from evaluations or fitness-report drafts. Nevertheless, think twice before taking phrases from elsewhere and planting them in an argument without adaptation. It can be dishonest in some cases, depending on how you represent the work. In other cases, using the material may be ethical, but it may have been written for a completely different audience and situation and thus may come across as out of place or confusing.

ADDITIONAL REFERENCES AND RESOURCES FOR EFFECTIVE WRITING

Books about writing style, grammar, and punctuation abound. The following resources are just a few of many and apply to principles of writing, proofreading, and editing in general. We list additional resources—including Navy- or Marine

Corps–specific ones—at the end of some chapters as well, applicable to the types of writing covered. Appendix A offers a complete list of the resources mentioned throughout this book.

Appendix B includes the *U.S. Navy Style Guide*, which provides Navy-specific usage examples as well as exceptions to the *GPO Style Manual*.

Books

The Best Punctuation Book, Period: A Comprehensive Guide for Every Writer, Editor, Student, and Businessperson, by June Casagrande (Random House, 2014).

The Elements of Style, 4th ed., by William Strunk and E. B. White (Pearson, 1999).

Garner's Modern English Usage, 5th ed., by Bryan A. Garner (Oxford University Press, 2022).

Department of the Navy Correspondence Manual, SECNAV M-5216.5 (2018; also used by the Marine Corps).

U.S. Government Publishing Office (GPO) Style Manual (GPO, 2016).

Online Resources

Grammar Girl: Quick and Dirty Tips, by Mignon Fogarty (https://www.quickand dirtytips.com/grammar-girl/). You can search for grammar and usage advice by topic or issue. Fogarty's explanations are clear, accurate, and logical, taking situation or varying usage into account.

Purdue Online Writing Lab (OWL), offered by Purdue University's College of Liberal Arts (https://owl.purdue.edu/owl/). You don't have to be a student at Purdue to access this enormous free resource of guidance on many types of writing and writing issues.

Plainlanguage.gov. Maintained by the Plain Language Action and Information Network (PLAIN), a U.S. government community of practice. This site offers guidance for writing clearly (in plain language) for government agencies.

Figure 1.4 provides the original famous quotations mentioned earlier.

FIGURE 1.4. *Famous Phrases Key*
1. I have not yet begun to fight.
2. Damn the torpedoes; . . . Go ahead, Jouett—Full speed.
3. You may fire when ready, Gridley.
4. I came, I saw, I conquered.
5. Don't fire until you see the whites of their eyes.

2

Naval Correspondence

"Tell young ensigns: 'Get to the ship, and take note of other people's correspondence. Figure out who's doing it right, and who's not.'"

—XO, NROTC UNIT

EXECUTIVE SUMMARY

A major part of most Navy and Marine Corps jobs is to send and receive mission-related communications. The three primary areas of interpersonal communication in the Navy are email, memos, and letters. All naval correspondence, regardless of type, should consider audience and purpose. Email provides many advantages but comes with several pitfalls that the naval writer can avoid by following best practices. Memos are for internal communication, while letters are for external recipients. Like email, both can be personal or impersonal, formal or informal. Most memos and letters fall into several types and will be most successful if writers follow time-tested principles for writing them effectively. The guidance in this chapter supplements and expands on the *Department of the Navy Correspondence Manual* (SECNAV M-5216.5). That longstanding reference (henceforth referred to as *Correspondence Manual*), used by both the Navy and the Marine Corps and readily available online, is essential for official guidance about formats and administrative processes related to correspondence.

EMAIL

"There is ABSOLUTELY NO PRIVACY when using email. Despite any admonishments otherwise, the 'FORWARD' button is easy to use—and once used, the damage is done, which no amount of correction, reprisal, or sanction can 'UNDO.' My advice is, if you wouldn't want it read by your mother/sister/boss/or the Washington Post, *don't use email."*

—NAVY CAPTAIN, PENTAGON

Since about the mid-1990s, email has become the most widely used means of electronic communication in the workplace. Few now on active duty will recall a Navy or Marine Corps without email. There is little question about its usefulness as a tool for workplace communication. Even though other forms—such as texting, cloud collaboration, and social media—have become popular for some tasks, email is still the primary way individuals and offices exchange information.

EMAIL ADVANTAGES

Just about every other document discussed in this book (for example, letters, memos, instructions, and award nominations) can be sent to its intended audience via email. These great advantages have led to its widespread use:

Ease of drafting. Email applications and electronic address books make composing and sending messages far easier than older methods of transmitting your thoughts to others.

Speed of transmission. The nearly immediate ability for your email to reach its recipient's inbox (compared to traditional "snail" mail, interoffice hand-carried memos, walking down the hall to talk face to face, or even calling on the phone) makes it an attractive medium of communication.

Ease of access. Nearly everyone in a modern work environment has immediate access to work email 24/7.

Security and reliability. Email provides a predictable, relatively safe means of information transmission, especially with the security protocols governing the Navy and Marine Corps email systems.

These advantages are so well known that they almost do not bear mentioning. We do mention them, however, for two reasons: First, our work culture has become so used to these advantages that we build our work flow around them; we expect all of these benefits, yet sometimes we cannot count on them. Second, some of these very advantages come with corresponding disadvantages and pitfalls we need to be aware of, as described in the next section.

EMAIL PITFALLS

"Good manners precludes using ALL CAPS AS A ROUTINE JUST BECAUSE ONE THINKS EMAIL IS JUST LIKE A NAVAL MESSAGE. No one likes being shouted at."

—PENTAGON STAFFER

Email's usefulness also comes with drawbacks. Although these will be familiar to most, they bear a little more detailed explanation than the advantages above. Let the list below serve as cautionary guidance.

Inappropriate Tone

"You can't pick up tone of voice in an email. You're never quite sure whether the written word doesn't perhaps imply sarcasm. If you have worries about your tone, use the phone or a conference call, or go see the person rather than use email."
—NAVY CAPTAIN

Email is an official mode of communicating in the military, the federal government, and the private sector, yet it also bears elements of personal, informal conversation. Perhaps the most frequent problem I hear of when giving training on email writing is that people misinterpret the tone of these messages, often hearing them as more angry or harsh than the writer intended. It is rare to find an official memo, a directive, or a set of instructions that comes across as demeaning, angry, or satirical. Yet these problems are common with email. The association of email with personal conversation makes it easy to write in a tone that comes across much harsher in writing than it might if spoken aloud in person. In commenting on this common problem, a chief of staff shook his head and grimaced: "You just hit the send button—and you can't get it back!"

The superintendent of the Air Force Academy found out about this lack of control some years ago when a "doolie" (a freshman cadet) wrote him to complain about the general's recent relief of a colonel on his staff—the young man had admired the colonel and was speaking in his support. In his reply the superintendent seemed to be offended that any fourth-class cadet would dare contact him at all, writing in a most pejorative and demeaning way to the clueless young student. Although the doolie perhaps should have known better than to write personally to the flag officer, the student's original email sounded a lot more professional than the general's overly harsh response. This fact must have been rather embarrassing to the superintendent and his staff, especially because this unpleasant exchange was forwarded from one person to another and (even in the days before social media) eventually found its way to the Internet.

Ease of Misreading

This problem can take several forms. The first is simply to lash out over email at somebody without forethought—and perhaps thereby make a permanent enemy. But it's just as easy to make a remark that might be interpreted as satirical or pejorative without the writer intending any such thing. "Sixty to seventy percent of communication is in body language," noted the CO of an NROTC unit when reflecting about this. The colonel's point is clear: in person, a joke, exaggeration, or well-meant satire is usually recognized and understood. But email is not in person and can't rely on physical cues. Along the same line, a Navy commander pointed out that emails are typically impersonal but often are taken personally. "And now, if you offend somebody, that person not only is offended, but they also have documentation of the offense!"

Ease of Miswriting

Others think the problem is often less in mistaken impression and more in the writer's original meaning. "People write things in email they would never say to anybody!" an XO exclaimed. Once again, this is probably attributable to the ease of immediate email communication—and the absence of the deterring effect of having your listener physically standing in front of you. Hence the importance of "taking a deep breath," as a senior chief yeoman put it: "Be as dispassionate in emails as you are in regular naval correspondence." A naval-writing expert recommended not addressing email drafts while you're working on them (or erasing the address you've already typed in or ported over, if a response). This way, a message can simmer in the draft file until you have time to think twice, and you won't inadvertently send it out with the touch of a key.

Informality

A former staff officer argued that "with a message or letter, you really thought about what you were going to say. You were careful with the pencil. But email is as informal as a phone call. And as with a phone call, there tends to be lots and lots of chit-chat." One result is that the official can mix with the unofficial. Because humor can be misunderstood and tone can offend, one command counsels, "Never say anything in an email that you would not say in any written correspondence OR that you would not say to your MOTHER!!"

Sloppiness with Language

Because we often associate email as an informal conversation rather than a formal document that requires thinking carefully about context, audience, and purpose, we are often not as careful about correctness as we should be. One chief of staff sent an informal email directive pointing out that staff work had too often been careless and that, in the future, documents for the commander's signature should be correct in all respects. Unfortunately for the O-6, the two-paragraph email message in which he counseled correctness contained no fewer than twenty grammatical errors. Subordinates were not impressed.

Email Overload

Throughout the service, officers, enlisted people, and civilians expend valuable time attempting to cope with a seemingly ever-increasing volume of emails.

"I left for seven days, and when I returned, there were 971 emails in my inbox!" commented the chief of staff of a midlevel command—this was back in 2005. Nearly two more decades of email use have not reduced the amount of email traffic we juggle. The XO of a small naval base reported that he and the CO sorted twice through the hundreds of emails received each day, just to make sure they didn't miss anything important.

We sometimes compare the informal system we all use to sort email messages to "triage," the process used by emergency medical technicians to differentiate between those patients who can wait for treatment and those who cannot. "The first thing we all go to when we arrive in the morning is emails—to see what we've been tasked with," a naval aviator pointed out. "What are the hot taskers?"

Some years ago a young man applied to the Marine Corps Officer Candidate program. He sent in his application weeks before the deadline, but a difficulty with an eyesight waiver caused the medical tests and paperwork to be forwarded to the Navy Bureau of Medicine and Surgery (BUMED) for a decision. The date for the selection board to meet came and went, with no word from BUMED. Upon inquiry, senior Marine Corps officers searched and discovered that BUMED's approval message was sent well before the deadline to a computer in the relevant office at Headquarters, Marine Corps (HQMC). But no one had thought to look for the message on that official's email. Meanwhile, although the young man's credentials were otherwise excellent, his file was incomplete, and consequently, he was not selected.

To prevent such problems, the *Correspondence Manual* directs that activities must spell out how to access and process email that comes to users who are absent for any significant period. Such procedures might include automatic forwarding of emails to other users or sending automatic-reply messages to email originators informing them that the addressee is not on hand (and indicating how to contact somebody else). Prudent commanders will regularly review and enforce such procedures.

Lack of a Paperwork Trail

In staffing correspondence forwarded in the traditional way (by hardcopy folders), everyone in a lineal chop chain sees criticism by other offices and understands why changes are suggested. Even when revisions are finalized and the folder is sent forward for signature, written concurrences and "nonconcurs" are automatically included in the file.

But in many cases, email critiques of staff documents are not printed out or even kept. As a result, officials reviewing staff work sometimes have little understanding as to why a final document appears as it does.

Email Tasking and Ambiguous Responsibility

Tasking by email is fraught with problems. In the past a CO could double-check whether the command was answering all requirements easily: One simply glanced through the message file in radio and the tickler file in ship's office. It's not so simple now. Frequently, a command will be tasked via email (by an office or code in a superior command) without the CO's even knowing. One commander found that some of his subordinates were turning down a headquarters request without his ever hearing of it—had he known about it (and thought it important enough), he might have been able to change their priorities so each responded "Yes." He directed that his people thereafter "never say 'No' without my say so."

To what extent can email be used for tasking, and if it is used for tasking, what problems can this pose? Offices within major commands now routinely task equivalent but subordinate offices in the field by email, rather than tasking the field commanders. Not too many years ago, naval commanders, to understand what their ships or units were required to do, followed the message board closely and paid close attention to the serial file in a ship's office or in the administration building to see what their particular ship or station was tasked to do. This practice enabled them to carefully track compliance. Now, however, fewer messages are sent, and letters often skirt the ship's office entirely—again, usually because they are attached to an email sent to a code within a command rather than sent via snail mail to the ship or the command itself (although paper letters often follow by regular mail).

Finally, email is a poor substitute for the personal. As one commander lamented, "With email we tend to lose the face-to-face." This comment was echoed throughout naval staffs. Naval men and women should be leaders, of course, not computer junkies. Know when to pick up the phone, schedule a video chat, or better yet, walk down the passage and have a cup of coffee with your shipmate.

Unwise Forwarding of Emails

Of course, there's at least one other related pitfall involving unwise speech and email. As a master chief advised: "Watch casual conversation in emails. Once you hit the Send button, you have no control over where your document goes, who sees it, or how it will be interpreted."

A Navy chief of staff recently advised, "Always know who is on your email's distribution list." He had once been sarcastic in a message. Thinking back on the situation, he still thought the sarcasm was justified—given the circumstances—because the command had suffered much from someone's error. But after sending the email, he realized that in addition to addressing his primary audiences, because of the structure of the distribution list, he had inadvertently sent the email "for information" to the commander of the individual who had made the original error—something he never intended to do.

Similarly, the XO of an NROTC unit insisted that people look at what they're forwarding before they forward it. She pointed out that using "Reply" does not necessarily involve sending all the attachments in an email, but the command "Forward" does. Hence, your comment to someone in an email chain or attachment—"Nope. We don't want to tell him that" or "Why is the XO sending us this stuff? It's all worthless!"—could easily embarrass you. When interviewed, that same NROTC XO (a veteran of tours in the Pentagon) reported that she had seen pejorative, thoughtless comments in attachments or email chains at the highest levels of the Navy.

FIGURE 2.1. *Brief Tips from the Author on Email Writing in Email Format*

5 MARCH 24

From: LCDR Chip Crane
To: Navy and Marine Corps Writers
Subj: EFFECTIVE EMAIL WRITING—QUICK TOP 10 LIST OF PRINCIPLES

I am writing to provide you with a brief summary of principles for writing emails that will be clearer for readers and more likely to achieve the purpose for which you send them.

1. Think about what your addresses, your readers, really need to know, and tailor the content accordingly.

2. Make your subject line as specific as possible without filling up the whole line.

3. Put your central point, your bottom line, as near to the beginning of your email as possible.

4. Organize the rest of your information to put emphasis on the most pressing or important information.

5. Use white space, bullets, numbered lists, indentation, short paragraphs, and other visual formatting techniques to make the key information stand out easily.

6. Make your expected or desired response (like "Please reply with . . ."), due dates, and other actions explicit and easy to notice in the email layout.

7. Use personal pronouns like "you" and "I" for most emails, and decide on the appropriate level of formality.

8. Adopt a friendly, professional tone, but keep annoyance, irritation, and other forms of anger out of your tone.

9. Summarize attachments so readers know which ones they need to look at carefully.

10. Proofread carefully, especially when writing emails on your phone. For important, sensitive, or complicated emails, let your draft sit for some hours—or even a day—before reviewing, revising, and hitting Send.

You'll find more details about these principles of email and about other forms of naval correspondence in chapter 2 of *The Naval Institute's Guide to Naval Writing, 4th Edition*. The *Correspondence Manual* (SECNAV M-5216.5) also provides some official guidance on the use of email.

V/r,
C. E. Crane

CRAFTING AN EFFECTIVE EMAIL

With some of these common email problems and pitfalls in mind, below are a few fundamental and practical qualities of emails that avoid those pitfalls and succeed in their purpose.

The sample email in figure 2.1 provides an overview of the principles for writing emails well. Examine this before moving on as we take a closer look at some of these principles.

Make Your Subject Line Specific

Making subject lines genuinely informative is vital to helping your reader know whether and when to open your email.

Upon reading dozens or even more than a hundred subject lines at a single review, people inevitably guess at the meaning of some emails and misinterpret others. Subject lines that do not introduce the topic in enough detail or that identify it poorly mean that readers spend extra time opening and reading messages that do not affect them—while missing others that do. As the minutes mount up, so can the frustration.

Therefore, make your subject line specific. One commander argued, "If you've managed to get your new ship's required manning [SMD] even further reduced, say in your subject line, 'As of 20 Mar 22 DD-XXX SMD down from 270 to 247.' For those who recognize the significance, that will catch their eye." In fact, he also pointed out that sometimes a knowledgeable reader will not need to read the email itself if the subject line is especially well crafted. You can even add "SLO" for "Subject Line Only" to your subject line and then leave the body of the email blank.

Also, remember to change the subject line when you change the subject. Often in a string of replies, the topic of conversation can change substantially. An email with a subject line "Next week's change of command" might produce some replies that end up discussing "Fitness reports," including some valuable guidance on the reports. Yet later, if you go back to search for that fitness-report information, you may forget that it was buried in an email chain with "Next week's change of command" as the subject line.

Make your subject line specific enough so that the people who need to read it will know to open it and understand what to expect. Consider which of the following options for the same email would be more helpful to the reader in deciding whether to open the message:

Personnel Reassignments

Assignment to Deck Division

Reassigning You as Deck Division Officer

Clearly, the third one gives the most information and will draw attention, whereas the reader (a reassigned JO) may assume from the first two that the email contains only routine admin that she may not need to read right away.

BLUF: Place the "Bottom Line Up Front"

As noted elsewhere in this text, for workplace writing in general you should directly state the main idea in your document up front as early as you reasonably can. This principle, BLUF, is even more important with emails than with letters and memos.

There are two reasons for BLUF. First, whenever an overly busy officer, enlisted person, or naval civilian actually opens your email while reviewing a long list of them, he or she is still looking to do "triage" to find out quickly what the point of the message is and whether it must be attended to quickly. Help your reader: get the key information into the first paragraph—even the first sentence, if possible.

A lieutenant teaching at the Naval Academy pointed out that this can be import-ant even when sending an email to your own boss. "Your XO may have asked you at morning quarters for your report but that afternoon may have forgotten what he asked you for! And he won't know what your report is unless you remind him at the beginning." A quick summary in the first lines (in addition to a good subject line) informs the reader of the document's point or contents.

Another reason to put important information up front in an email has to do with the "preview pane," or the quick view of the beginning of a message that many email apps provide.

A principle in journalism is to get your main idea "above the fold." That advice refers to the fact that traditional newspapers originally were folded in half, and a story (or at least a headline) above the fold meant readers were much more likely to see and read it. In addition, for news releases, journalists have always been instructed to put the most important information—the who, what, when, where, why, and how—at the beginning of the story (in what is called a "news lede"), thereby enticing readers to continue down the column and ultimately to turn to the interior or back pages of the paper to complete the story. Exercising the BLUF principle will ideally give your reader enough information to decide whether to open the email and read on—or, upon opening it, whether to scroll down the page to glance at material not immediately visible.

Organize for Clarity and Emphasis

After your BLUF, tell readers what kind of structure to expect, especially in a longer, more complicated email. If you are going to cover three points or describe a prob-lem and then a solution, include a sentence telling them to expect that.

We also recommend breaking your main ideas into small paragraphs—smaller than you might in another type of document—and providing frequent headings for the main topics covered. Doing so may feel like overkill in a less formal email, but remember how much skimming we do, especially when reading messages on our phones. Extra white space and headings will go a long way in helping your reader catch the most important information in your emails.

Briefly Summarize Attachments

Opening attachments also requires time, so it helps your readers to know what is in them so they can decide which ones to open first and identify those they can read later. An example of such a summary would be, "Attached you will find our proposed ship requirements for the XXX-Class Ship." You might need a short paragraph to summarize a complicated document, however, not just a sentence. The CO of a training command pointed out that such summaries are not just "nice to have" features but are now more and more expected.

Emphasize Action

Throughout your email, make clear any action people need to take. For example, consider the following fictitious email.

DRAFT

LT Edwards,

In light of last week's meeting with the department heads, a number of actions are in process and on track. The evaluations and counseling for the Sailors in my division will be completed in the next two weeks in accordance with the schedule promulgated by the XO. In addition, a review of the requirements for the change of command will be used to put together an action plan by the end of this month. The leave requests for leave prior to the underway period in April should be submitted in the next several days as will any concerns or exceptions with those. Overall morale in the division is less of a problem than it was.

V/r,
ENS Gibbs

This email buries important information in a paragraph covering a variety of topics, making it difficult to focus on each issue. Furthermore, it obscures who is doing what with unnecessary passive voice. Who is planning to do what and who is responsible for each remains unclear. And a quick skim through the email will likely result in missing some details. Consider how the revision below rewords sentences and restructures the paragraph to emphasize action and remove ambiguity.

REVISION

LT Edwards,

At last week's department head meeting, which I attended in your place while you were on leave, the XO emphasized several important deadlines for our department. I am on track with all of them as follows:

- I will submit evaluations and mid-year counseling for the Sailors in my division to you by April 2 per the XO's schedule.

- I will review the requirements for the change of command and give you an action plan by the end of this month.

- All of the Sailors need to submit their leave requests for leave prior to the April underway period by March 23. They also need to explain any special circumstances and submit those along with their leave requests.

Also, you will be interested to know that morale in the division has improved in response to the new policy you implemented last month.

V/r,
ENS Gibbs

Choose and Use an Appropriate Tone and Style

In many writing contexts, setting the appropriate tone can be difficult, but email may be the area in which doing so challenges people the most. Because email seems to be a simple form of communication, sometimes more of a conversation, writers often take less care to really think through their tone and style than they do when writing a letter or a new policy. In addition, the semiconversational nature of many emails often clashes with the semiformal (or formal) context for which people use it.

Formal vs. informal. This choice is not an either-or but rather one of degree on a sliding scale. Will you write that the command needs to "determine" or "figure out" the best course of action? Will you write "do not" and "they are" or use contractions like "don't" or "they're"? Either option can work well in the appropriate context. Small word choices like these help set the tone of an email (or any communication), just as wearing jeans to an event creates a less formal mood than wearing a suit or a dress uniform.

Impersonal vs. personal. This choice is also closely related to tone, and you establish it largely through the use of personal pronouns. As discussed regarding plain language in chapter 1, addressing the reader directly as "you" and saying "I" or "we" creates a sense of people communicating with each other. Writing in the third person about "personnel," "the office," "the command," and so forth emphasizes the roles and the institution. For some documents, the more impersonal tone is more appropriate; for many emails, however, a more personal tone works best. Once again, the audience and situation should influence these choices.

Emotional vs. objective? As professionals of whatever rank, we expect from others—and pride ourselves on—keeping workplace communications, including emails, focused on professional, objective information. Yet as humans, we also feel excited, angry, sad, betrayed, exuberant, and many other emotions. Sometimes, including some enthusiasm in an email of praise or encouragement is appropriate, but you should do so on purpose rather than unintentionally. Negative emotions like anger (or a milder form such as irritation or annoyance) are almost never appropriate for an email. But it is difficult to keep them completely out of your tone if you are feeling even slightly annoyed. A Navy supply officer and former writing center director has noted: "Writing tends to magnify emotional effects to the reader. If you write, 'You need to submit those reports to me by 1600 or the division has to work late,' the mild sternness you might imagine you are conveying could easily come across unintentionally as 'TURN IN THOSE REPORTS BY 1600 OR ELSE!'"

Use Appropriate Closings

In email and other correspondence, remember to use standard naval protocol for the complimentary close of documents sent to seniors or juniors:

"Very respectfully," used by juniors writing to seniors

"Respectfully," used by seniors writing to juniors

In email you can shorten these as follows:

"V/r," (juniors to seniors)

"R/," (seniors to juniors)

When writing to peers, these are not required, but it is a common additional courtesy to use one of the "very respectfully" forms. (Someone commented in an interview that Marines do not usually make use of such penned salutes; instead, they typically end their emails with "Semper Fi.") As one Naval Academy instructor of professional writing advised, "Use the same rules of military courtesy as you would in naval correspondence."

Your email signature block should typically include at a minimum your name, rank, command, email address, and work phone. Different commands may have specific guidelines for signature blocks, so check with your administrative office.

EMAIL ADMINISTRATION

Just as with the guidance about composing emails, what follows are some administrative issues to consider with managing email. Your command may have specific guidance covering some of these things, but most of this will depend on common sense and situational awareness. The advice below is based more on some best practices observed in the fleet, but it is not meant to be overly prescriptive.

Decide on Whether to Keep an Email Chain

When interviewed, officials pointed out reasons to keep—and reasons to expunge—an email chain (earlier messages that have led to the current one). A Navy captain reported his regular frustration at spending ten minutes with a string of emails, each with its own attachment—only to find that the key information was in the last attachment. Given such difficulties, consider summarizing the negotiations from all those emails and deleting the chain, especially if one's document is to be sent to multiple addressees or if it is setting policy based on a long chain of such messages.

On the other hand, when coordinating staff work, such email chains serve the purpose of letting everybody know how the negotiations have gone on a particular issue; that is, which offices have pointed out specific problems (or specific solutions) and their reasons.

Filing Emails

Email is also useful for recordkeeping. Unlike phone calls, email provides a first-hand written record of who said what in a conversation. Keeping such conversations can protect yourself, in some cases, or can clarify who is responsible for specific actions. Keeping them can also be useful for historical use—you might want to look back in several months to help remember how something was worded or what was put out. A sequence of emails can also be valuable in debriefing an exercise or creating a lesson learned for future use.

There may also be situations when you do *not* want to keep emails—the issues of whether to include email chains and whether to retain them can be complicated. See in relation to this topic the discussion of keeping message references (found in chapter 5) and the section on staff coordination (in chapter 3).

How "Official" Is Email?

When asked about our use of emails, a Navy chief of staff reflected on the breakdown of the traditional protocol that used to govern letters and memos. Traditionally, a memo has been for "internal" readers, while a letter was for "external" audiences; that is, a letter is typically always sent from one command to another (or sometimes from a command to a service member or vice versa), but not a memo. Emails are a different kind of document altogether, possessing qualities of memos, letters, and phone calls all at once. While they often look like memos, emails are sent externally and often from an individual within one command to an individual within another. They also can be informal in style and tone.

This same chief of staff reflected on the problems caused by the breakdown of this protocol. He commented that when we used to write a letter, we took care with it, recognizing that even if it was signed "by direction," it still was understood as a representation of the command. In his view we should regard our emails to be just as representative of the command as our letters.

Along these lines, a Navy commander said that in her judgment the main issue is to what extent emails are "official." Certainly, the more official one's job, the more one is expected to write an email in official memo format, with the exception of not numbering the paragraphs with emails (as one does with memos, discussed later in this chapter). In this view one should always treat an email as official.

A related issue is the ease with which subordinates can end up making command decisions on informal emails. A master chief yeoman argued that it is important that senior enlisted administrators not make command decisions via email. She reported that yeoman master chiefs in particular are notorious for speaking for the command without authority or for answering a query and only afterward researching an issue—and then having to back down.

And a commander mentioned this difficulty: "One lieutenant at a shore command answers a question of his counterpart lieutenant (or lieutenant commander) at his governing command. He's thereby sent out official correspondence. Often, that lieutenant should check with his chain of command first." In his opinion, with email just as with paper correspondence, there's always a proper "releasing authority."

The point is that often your boss or others in the command may have a specific interest in what you've been asked and how you're responding. You should keep that in mind and keep your superior—and whoever else may need a say—in mind. "Otherwise, you'll learn by experience," a commander concluded. "That is, you'll learn by screwing something up."

Use the Tools

It is important to learn something of electronic "office management" systems. Take the email distribution system, for instance. There are lots of groupings possible—emails to be sent to "All COs on base," to "All Master Chiefs," and so on—within a command. Emails also have many possible distributions: to the whole command, to each code, to all officers, and so on. Knowing this, you can reduce your workload by adjusting your profile so that you are included in the groups you should be included in to avoid receiving emails that you don't need to read. A chief of staff at Naval Reserve headquarters found that until he got the distribution list changed, for some unfathomable reason, he was getting dozens of "Explosive Impact" reports. Also, frequently after you have moved to a new position, email still comes from connections to your old job until you let people know you've moved on.

Mail management systems—such as Microsoft Outlook—offer tools to make your life easier. You can usually set up a filter or "rules" grouping the emails you should read first—perhaps "everything from your CO" to start with. Then a CO can set up the system so everything he sees also goes to the XO, or whatever comes to the XO can also be sent to the command master chief.

As with Microsoft Word, you can set up your email to be automatically spell-checked. One CO said he always did this, because who knows what spelling errors one might commit in a quick email response and then where that response will end up.

Memos

Memoranda range from brief notes to vital policy initiatives, and of course, many emails are essentially memos. They are mainly for internal use aboard staffs and operational commands but can occasionally be used externally between commands. You would not, however, send a memo to someone outside the Navy—you would send a letter (see the section below on "Letters" for corresponding with external audiences). Most memos, even when following a traditional or command-prescribed format, will still be circulated via email, either as attachments or in the body of the message. A formatted memo is generally seen as more formal than an email, but both are official, legal correspondence.

For greater detail on formatting and the varying uses of memos, see the *Correspondence Manual* (SECNAV M-5216.5).

TYPES OF MEMOS

The *Correspondence Manual* identifies six types of memos for regular communication, each for use in different situations:

Memorandum for the Record

From-To Memorandum

Plain-Paper Memorandum

Letterhead Memorandum

Decision Memorandum

Memorandum of Agreement, also called a Memorandum of Understanding

A few additional types of memos fall under the category of "executive correspondence." Memos from Navy headquarters (OPNAV) have particular usage and formatting requirements due to the weighty import and formality when communicating with the Office of the Secretary of Defense, the White House, and other federal departments' secretaries as well as with units within the Navy. These executive memos are

Action Memorandum

Information Memorandum

Memorandum For

The following subsections provide examples and some discussion of several formats and uses of memos, especially the common ones. Keep in mind that each command may have different practices and formats for memos, so you should check the advice here against what your command prefers. The bulk of the discussion on writing applies to all types of memos, but we will briefly explain some of the different contexts or approaches for each kind. We spend more time on a few memo types than on others. Consult the *Correspondence Manual* and your own command admin office for details about formatting and routing requirements for all memos.

Memorandum for the Record

Do you want to ensure some important information is recorded but are afraid that, because of the informal circumstances in which it came up, it won't be? Consider using a memorandum for the record (MFR). This is a memo by which you record details about a situation that might otherwise be forgotten and lost. It has wide potential usefulness. You are putting together some information you want to save for later use, and making it an MFR gives it a more official, clear status than a set of loose notes you might save with a vague filename.

The information in an MFR might be from a meeting, a telephone conversation, or an informal discussion held on a staff. You can use it to create a record of an agreement among several parties at a conference or of decisions made at decision briefings. The MFR resembles the minutes of a meeting in some respects. Another common purpose is for documenting information from a phone call or from an informal discussion that an investigator may conduct in the process of a JAGMAN investigation.

An MFR will not have a "to" line because it is not for a specific individual or group. This does not mean you should ignore considerations of audience when writing it. Think about who might use the MFR in the future and include enough information so that it will make sense to anyone (perhaps even your future self, after you have forgotten most of the details) who consults it months or years from the present.

Usually, a person files an MFR for future reference, but you can route it to your staff if everyone needs to know the information it contains. Staff officers can forward them for record up the chain to keep seniors informed of what's happening down below. Whatever you use it for, keep this memo informal. It is an in-house document to help keep track of business. Do remember to sign and date it, but always keep it easy to use. Most likely, you will file it on your computer and the command servers, so choose a file name and folder that will make it easy to retrieve when needed.

The example MFR in figure 2.2 is fictional, but it is based on some used as exhibits in a JAGMAN investigation. It documents a discussion between the investigating officer and an expert about funds that had been stolen from a postal safe.

FIGURE 2.2. *Memorandum for the Record (Conversation)*

July 2022

MEMORANDUM FOR THE RECORD

From: LT J. R. Black, USNR, Investigating Officer

Subj: RESPONSIBILITY FOR CHANGE OF POSTAL SAFE COMBINATIONS ON USS OVERHAUL (FFG 999)

1. On 9 July 2022, I discussed this investigation with PCC Gray of the COMCENTGULF Postal Assist Team. Specifically, I asked PCC Gray what the responsibility of the postal officer in this case would have been. He said that a postal officer must oversee the entire postal operation of the command. Therefore, ENS Brown did have a duty to make sure safe combinations were changed. However, he also pointed out that ENS Brown's responsibility was oversight only, and that the primary responsibility for changing the combinations remained that of the Custodian of Postal Effects (COPE) aboard USS OVERHAUL, that is, PC2 White.

J. R. Black

FIGURE 2.3. *Memorandum for the Record (Information)*

15 May 21

MEMORANDUM FOR THE RECORD
Subj: TARGET ANCHORS AND SALVO RETRIEVERS

1. I have been asked the following questions by OP-88Z and by Mr. A. C. E. Shooter of SASC staff:

    a. What are the quantity and funding profiles for target anchors and salvo retrievers for FY 19 and prior through FY 21?

    b. What would be the cost of 300 target anchors in FY 21?

2. I provided the following information:

	Target Anchors	Salvo Retrievers
FY 19 & Prior	150/$3.0M	250/$15M
FY 20	300/$5.4M	400/$23M
FY 21	400/$7M	550/$32M
FY 22	50/$0.5M	800/$46M

The cost of 300 target anchors in FY 21 would be $4.8 million. The inventory objective for target anchors remains 1,500; the inventory objective for salvo retrievers is 3,000.

3. This inquiry is probably the first of many on this subject. We should be consistent in our answers.

                                          E. Preble
                                          Director, Targeting Division

Copy to:
OP-OX

Specifically, the officer wants to know who was assigned the responsibility for the safe's security.

Clearly, it is important that such a discussion as illustrated in figure 2.2 be recorded right after the discussion takes place, and the MFR provides a good way of doing that.

The example in figure 2.3 is somewhat more formal than the previous one. It resembles an MFR that might be put out by an office in OPNAV or at another major staff. Rather than documenting information discovered in an investigation, here a division director sends cost figures on two major equipment procurements to individuals in the chain of command and to members of his own staff.

From-To Memorandum

The from-to memorandum is the most common and arguably the most important. As the name suggests, this memo includes a "From" creator and a "To" audience, often a group. Figure 2.4 is a from-to memo on a routine matter written in an informal

FIGURE 2.4. *From-To Memo*

20 February 2024

MEMORANDUM

From:　Admin Officer
To:　　Department Heads
Subj:　LETTERS OF DESIGNATION

1. There are numerous requirements laid on by this command and higher authorities to "designate in writing" individuals to perform certain tasks/responsibilities/accountabilities. It would behoove us to know exactly who all these folks are and "track 'em."

2. I'm requesting that each department review its working instructions, CVWR-30, and higher instructions (such as 4790.2) and forward a list of all such requirements to Admin NLT Thursday, 29 February.

3. Admin will collate your responses into a notice, and we will thus be assured that all requirements to designate are met.

P. D. Brady

---

style. As noted above, most people now would use an email to say the same thing, but the principles of structure and clarity are the same. Print this type of memo on plain white paper (that is, no letterhead) or create a PDF in the same format.

Plain-Paper Memorandum

In the past a memorandum typed or handwritten onto a preprinted form or on plain paper was the least formal type of official written communication. Although this kind of memo is still in use, it is far less common now because we normally use email—or in some cases a sticky note—to accomplish the same purposes. We mention it here primarily so you know what it is if someone mentions it.

Letterhead Memorandum

A letterhead memorandum, as the name suggests, has the command letterhead at the top, whether in print or in electronic form. The letterhead memo is more formal than the plain-paper memo, and it represents the voice of your command (that is, your CO's authority). You should not use letterhead for a memo (or even for a letter) when you are writing as yourself and not on behalf of the command or unit.

Decision Memorandum

Use a decision memorandum when the purpose of the memo is to provide a decision maker (most likely your CO) the necessary information about an issue she or he needs to address. The body of the memo will explain the issue and request or recommend approval of a course of action. Below the signature line, add a block of lines for the CO to sign, indicating his or her decision.

COMMANDING OFFICER DECISION:

_____ Approved

_____ Disapproved

_____ Other

As with the other memo and letter types discussed here, details about proper format appear in the *Correspondence Manual*.

FIGURE 2.5. *Memorandum Used for Performance Counseling*

31 March 2022

MEMORANDUM

From: Maintenance Officer
To: ADC Romeo
Subj: PERIODIC PERFORMANCE COUNSELING, 10 Oct 21–31 Mar 22

1. This counseling is being held to document your performance from 10 Oct 21 to 31 Mar 22. This is your first periodic counseling since joining the department. You have done an excellent job of integrating yourself into the Quality Assurance division. In addition to the numerous qualifications you already possess, you have progressed well toward additional qualifications.

2. Training. Your attention to detail and training competence are evident. Your actions played a vital role in preparations for and conduct of the successful March 04 CNAL AMMTI. As for the future:

   a. I have counseled ATCS Murphy to work closely with you so that you learn the techniques and methods that he uses in the Quality Assurance process. You should be able to perform the bulk of his missions within 6 months. Obviously, at least half the effort here will be yours.

   b. Continue to train OARs and Detachment CDOARs with the objective of having others perform tasks except those which you are specifically required or uniquely qualified to perform.

   c. Emphasize practical evaluations during DMPA and Weekly Maintenance training. Try to drive the training away from simple reading of written material. When able, personally oversee training for quality. I recommend you use CNAL AMMTI team technique where practical results drive the focus on program evaluation.

   d. As you are the AVGFE, I expect you to maintain a flawless Aviation Gas Free program that is inspection ready at any time.

   e. Evals: Safe for Flight within the next 30 days.

3. Leadership. As an individual who went from E-5 to Chief in three years, you can provide much needed leadership to the First Class Petty Officers in the department. Use and create opportunities to influence and instruct them.

4. You are a vital part of this department. You are an OAS in training and should see yourself that way. In addition, use your specific experience to lead the First Class POs.

                                                                A. N. Officer

---

FIGURE 2.6. *Guidance Example for Preparing a Memorandum For*

[LETTERHEAD]

5216
Ser 943D /345507 10 Jan 23
MEMORANDUM FOR THE DEPUTY CHIEF OF NAVAL OPERATIONS (OP-XX)

Subj: PROFESSIONAL PREPARATION OF THE MEMORANDUM-FOR INFORMATION MEMORANDUM
Ref: (a) CNO Supplement to DON *Correspondence Manual*
        (b) HOMC Supplement to DON *Correspondence Manual*

1. The memorandum-for is a formal memorandum. Its normal use is to communicate with senior officials such as the Secretary of Defense, the Secretary of the Navy, the Chief of Naval Operations, one of the Assistant Secretaries or Deputy Chiefs, or an Executive Assistant for any one of these officials.

2. Take great care in the preparation of a memorandum for. These documents have high visibility and require thorough staffing and tactful expression. Make sure each of them has the following:

   a. a subject line that best describes the memo's purpose;

   b. headings, if useful; and

   c. brevity, always: normally keep the memorandum to one page.

3. If you use tabs, be sure not to let those tabs substitute for good staffing. Do your best to pull the relevant information out of the references and weave it into your memorandum rather than asking a senior official to plow through the tabbed material.

4. Protocol is important. List the addressees in the established order of precedence.

5. Prepare the memorandum for letterhead stationery. Because it lacks a "From" line, show the signer's title below the typed name.

6. Various offices have issued additional guidance on preparing this type of document. For example, the Secretary of Defense once asked that "ACTION MEMORANDUM" or "INFORMATION MEMORANDUM" be placed at the end of the subject line of each memorandum for, and OPNAV offices have generally followed suit. See current versions of references (a) and (b) and other local information for up-to-date guidance.

                                              J. Memorandum
                                              Deputy Chief of Naval Operations

---

Memorandum of Understanding

The memorandum of understanding (MOU), also known as a memorandum of agreement (MOA), perhaps differs the most from all of the others. This memo is more like a contract between two commands or a command and another government agency. It can document an agreement on certain facts, procedures, roles in a partnership, transfer of funds, and other actions or topics. Chapter 10 of the *Correspondence Manual* details the elements of an MOU or MOA.

Another kind of memorandum, the briefing memo, is widely used on staffs in briefing folders. Chapter 3 in this book discusses the briefing memo as a type of document frequently used by staffs.

Memos can of course be used for many purposes. In figure 2.5 an officer uses the memo format to record a counseling session with a chief petty officer. In this well-written memo, the officer details specific directions on how the chief should proceed (implicitly setting the standards by which he will be evaluated). On that subject, a Letter of Instruction is another common, but more formal, tool for documenting counseling (see the "Letters" section below).

Memorandum For

As noted before, the memorandum for is a more formal type of memo common to OPNAV and other staffs. The example in figure 2.6 demonstrates the format; its content offers guidance for this type of memo itself.

LETTERS

"One of the most rewarding and successful parts of my command was, surprisingly, the form letters I sent to parents. I wrote each parent of the fifty people up for awards and promotions, modifying each letter slightly. The parents fired back letters to their kids, and morale zoomed up. It was a building block for success."

"Such letters don't have to be long, or sexy, or even grammatically correct, but they have a great payoff."

—NAVY COMMANDER

STANDARD NAVAL LETTER

Use a standard naval letter to write to organizations within the Department of Defense. Commands and individual service members write standard naval letters, the marks of which are a formal tone, the use of standard naval terminology and abbreviations, and a specific format. Use a business letter for writing to people and organizations outside DoD and also when the situation calls for a more personal approach, such as a letter of condolence. Official letters are, of course, often sent electronically as attachments to email or uploaded to a database or cloud server, yet they are distinct from emails. Despite the common use of email in many contexts that in years past would have required a letter, many situations in and outside of the armed services still require a proper letter, and you should work to master this longstanding professional document type.

As with emails and memos, the *Correspondence Manual* (SECNAV M-5216.5) is the authority on format and administration of Navy letters. It is readily and publicly available online. The advice below centers primarily on writing and usage guidance for some of the most common situations. Look to the *Correspondence Manual* for

FIGURE 2.7. *Guidance Example for Preparing Standard Naval Letters*

5216
N13
1 Mar 24

From: Authors, Guide to Naval Writing
To: Navy and Marine Corps Writers

Subj: COMPOSING THE STANDARD NAVAL LETTER
Ref :   (a) SECNAV M-5216.5, Dept. of the Navy *Correspondence Manual*

1. A standard naval letter should target the specific audience, purpose, and situation with the appropriate content, structure, and style and should follow the formatting guidelines of reference (a) and your specific command.

    a. Do not simply use another letter you can find from another Sailor or officer as a template—it may not be correct. Go to the *Correspondence Manual* for details of format and proper routing. Others' successful examples can be helpful for structure and style but only if you know they achieved their purpose.

    b. If you write a poorly formatted letter, expecting your admin office to fix it up, you should also expect administrative delays in processing your letter.

2. Here are some general guidelines for content and structure that apply to most letters:

    a. Bottom Line Up Front (BLUF). Put your chief request or central idea somewhere in the opening three or four lines to help get the attention of the right people from the very start. As reference (a) puts it, when you write a letter, think about the sentence you would keep if you could keep only one. Many letters are short and simple enough to have such a key sentence. It should appear by the end of the first paragraph. The strongest letter highlights the main point in a one-sentence paragraph at the beginning.

    b. Throughout the letter, put requests before justifications, answers before explanations, conclusions before discussions, summaries before details, and the general before the specific.

    c. After the opening, spell out the details. Write in relatively brief paragraphs—normally no more than four or five sentences apiece. Writing short paragraphs and punctuating them by white space makes reading easier; long paragraphs can discourage the reader and encourage skimming.

    d. Keep most letters to a single page. If you have several pages of material, use enclosures to spell out additional content. If your letter does need to be longer, use headings to keep the reader oriented and identify major topics.

    e. Letters, even though formal, are also personal by nature. Use "I," "we," and "you" naturally while maintaining a proper respect for the rank and position of your audience.

    f. As with nearly all reader-friendly writing, minimize passive voice, avoid wordiness, and follow other principles of "plain language" as described in chapter 1 of this guide.

FIGURE 2.7. *(continued)*

3. Follow this additional guidance for the other elements of your letter:

   a. Show codes and titles in addresses. Whenever practical, indicate the office that will act on your letter by including a code or person's title in parentheses right after the activity's name. If you are writing to a specific person, use the name and include the person's title afterward. If you are writing to an office, use the office.

   b. Compose a good subject line. Craft the subject line to make it genuinely informative. Try to limit it to 10 or fewer words. In a reply, normally make the subject line the same as that of the incoming letter.

   c. Get the other details right. See reference (a) for further information about standard-letter format, serial numbers, markings on classified letters, and joint letters.

   d. Always include your phone and fax numbers and email address when your correspondence might prompt a reply or inquiry, and make sure you include your own office code.

   e. Don't use a complimentary close ("Sincerely," etc.) on a standard naval letter. For rules on signatures (on who signs the letter, on "by direction" authority, on how to put together a signature block, etc.), see reference (a).

4. Determine the best means of communication and the appropriate form:

   a. Use the standard naval letter to correspond with DoD activities primarily, but also use it with the Coast Guard and some contractors.

   b. Send business letters to other external addressees.

   c. Before you even write the letter, make sure some other means won't suffice. Telephone calls, videoconferences, emails, and in-person meetings should be your first approach. You can always document these with a memo for the record if necessary.

5. Reply promptly. Answer most received correspondence within 10 days. Congressional inquiries require a response in 5 working days. FOIA requests have 20 days for a response (see reference (a), p. 26). If you don't anticipate being able to answer within the appropriate time frame, inform your correspondent of the delay and when you expect to reply.

C. E. Crane   R. Shenk

Copy to:
USS ALLHANDS (NAV 1)

---

details about format, routing, structure, signature authority, and other aspects of letter administration.

Consider the advice on how to compose the standard naval letter in figure 2.7, presented in Navy letter format.

OFFICIAL PERSONAL LETTER

The technical details of communicating with higher commands on personal matters (such as enlisted applications for warrant officer or limited duty officer [LDO], officer applications for augmentation or change of designator, and so on) are spelled out clearly in the *Correspondence Manual*. It recommends the following procedure for Navy personnel:

- Prepare your letter on plain bond paper (unless submitting electronically) in standard letter format.
- Address the letter to the higher authority and send it via your chain of command.
- Each "via" addressee prepares an endorsement and forwards the correspondence to the next addressee.

For individual requests, Marine Corps personnel use a special form (NAVMC 10274, Administrative Action [AA] Form) as prescribed in the Marine Corps Records Management Program (MCO 5210.2). Sometimes an office will authorize submission by email.

Normally, the format to use in personal letters is thoroughly specified by the responsible authority. In all blocks requiring precise data, the only way a person can go wrong is to leave something out, supply faulty information, or make errors in grammar or spelling. Any such mistakes might admittedly call into question the administrative ability of the applicant. If you pay attention, it's difficult for you to go wrong in supplying that information.

Getting the format and grammar right are only a small part of a successful letter. How well you write the body of the letter, the remarks section, can be crucial to its success with your intended audience.

Writing a Good Remarks Section

Usually, you craft your remarks to add key information and express the nature of your interest or desire. There are several possible approaches. One way is to recount, in brief summary form, what special qualifications or interests make you well suited for the position or new opportunity for which you are applying.

For instance, in the following excerpt, a surface warfare officer justifies her request for a change of designator to supply by emphasizing her strong academic background, her experience and course work in business, and her operational experience. She manages to work into the last paragraph many of the pertinent highlights of her naval service. True, these highlights might be picked up by board members in their review of her service record, but she makes sure the board sees them by mentioning them in the letter.

Having completed nearly four years of service as an Unrestricted Line Officer, I strongly desire to broaden my career by transfer to the Navy

Supply Corps. My civilian experience includes a strong academic background:

- B.A. (summa cum laude, Phi Beta Kappa), Univ. of Pennsylvania
- M.A., Cambridge University, England
- Several positions in the business and arts management fields, often involving considerable fiscal responsibility

Following graduation with honors from Surface Warfare Officer School, I began a tour of duty aboard USS *McCampbell* (DDG 85), holding demanding billets in engineering and operations. As both Auxiliaries Officer and Administrative Department Head, I have worked closely with the ship's Supply Department and have gained much insight into Supply procedures.

Additional details in the above could include the positions she held in business and arts management or perhaps the scope of fiscal responsibility she has had (for instance, "I was responsible for a $2M budget").

When applying for career opportunities, besides mentioning your own accomplishments as in the example above, you can also cite command accomplishments to which you've made a significant contribution. For example, consider the closing section of a reserve lieutenant's letter request for augmentation to the regular Navy, shown in figure 2.8. He thought his command's good maintenance record reflected

---

FIGURE 2.8. *Closing Segment of a Personal Letter Requesting Augment*

During my tenure as Maintenance Material Control Officer (MMCO), Fighter Squadron THIRTY-THREE has been recognized as one of the top commands in Fighter Wing One. Functioning during a post-deployment period with the lowest requisition priority in CFW-1, the oldest F-14A aircraft in the naval inventory, and substantial Fleet Maintenance Fund reduction, VF-33 has aggressively pursued operational commitments rivaling those of any deployed unit. Significant accomplishments while I have been MMCO include the following:

- 100% sortie completion rate during FFARP.
- Best missile expenditure record in fighter community for FY 1987.
- Overall FY 1987 sortie completion rate of 97.2% during 3657.6 flight hours, with mission capability rate high of 87.2% (CNAL average is 74.4%).
- Achievement of zero NMCS/PMCS requisitions on three occasions and maintenance of zero FOO rate during CY 1987.
- Lowest XRAY message error rate CY 1987 (highest in CY 1986), and finest aircraft logbooks in Fighter Wing One.

These achievements reflect my commitment to aviation maintenance duty. I am resolute in my career intentions and respectfully request augmentation to the regular Navy.

well on his own performance as maintenance officer and enhanced his request for augmentation.

This closing statement of the example spells out impressive specifics, accomplishments that stemmed from this officer's own work. He even supplies comparative data for the board's information, and his closing comments are forthright.

Templates and guidance for various personal-request letters, letters to the promotion board, and endorsements (discussed briefly below) are available at www.MyNavyHR.mil.

Endorsement to a Personal Request

The endorsement to a personal letter is much like a letter of recommendation and is usually freeform. A commander can write endorsements as a few short paragraphs or can use a bullet format (as adapted to standard naval letters). The endorsement does best when it begins with a summary statement (BLUF), then sketches specifics, and ends with a clear assertion of the recommendation. It can have a much stronger

---

FIGURE 2.9. *Endorsement to a Personal Request*

FIRST ENDORSEMENT on Gunnery Sergeant Frank L. Competent, USMC, 123-45-6789 of 28 Mar 22

From: Commanding Officer, U.S. Naval Construction Battalion ONE
To: Commandant of the Marine Corps (MMEA-86)
Via: (1) Commander, TWENTIETH Naval Construction Regiment
Subj: REQUEST HUMANITARIAN TRANSFER

1. Forwarded, most strongly recommending approval. Based on my review of this case, Gunnery Sergeant Competent's presence at home would be instrumental in resolving this hardship.

2. Gunnery Sergeant Competent has been working with his chain of command, the Command Chaplain, and Navy Family Counselors to assist in his family's challenges. Due to his strong will and dedication to the Battalion, he has refused to allow these difficulties to affect his superior performance as NMCB ONE's Marine Advisor. He has used his leave to attend appointments, surgeries, and physical therapy treatment. Gunnery Sergeant Competent and his son are truly stoic men, but the Competent family has endured many extreme tragedies, and granting him a humanitarian transfer will be a great benefit to his son's mental, physical, and emotional well-being. Furthermore, if allowed to remain in the local area, Gunnery Sergeant Competent 's son will maintain medical continuity, and he will greatly benefit from his current physician.

3. Granting this request would allow Gunnery Sergeant Competent the opportunity to give his family increased stability and emotional support. Additionally, the Gunnery Sergeant's superb ability to work with Seabees would greatly benefit the Navy and the TWENTIETH Naval Construction Regimental Marine Advisors. Since this Battalion will be deploying to Japan in June for six months, I recommend this humanitarian transfer take effect immediately.

4. Gunnery Sergeant Competent is not pending any disciplinary action.

R. L. Approver

effect than any single performance evaluation. Rather than being filed away in the service record with all the other fitness reports or evaluations, the recommendation remains attached to the request while it is being processed. Although the service record may be on hand for those who make the decisions, the command's recommendation will usually be the only evaluative document to comment on the specific request being made.

Clearly, it is important to craft the endorsement well. Not only commanders but also juniors who make requests should know how to write them. Juniors should normally submit a draft endorsement for their superior officer to sign along with the request letter.

What do you include in an endorsement or a recommendation? First, speak specifically to the particular request being made. Specific comments are relatively easy to make if the member's past service relates to the request at hand. The paragraph below endorses an unrestricted line officer's application for redesignation as an intelligence officer. Because the applicant has been the squadron's intelligence officer, the endorser can speak directly to the person's intelligence-related duties:

> He is a proven performer who has substantially enhanced the squadron's intelligence and radar identification training program. In his role as an intelligence instructor, he has provided positive, aggressive leadership and technical guidance during the intelligence/mission-planning phase of fleet replacement aircrew training. He has also performed expertly during the Medium Attack Tactical Employment School and during weekly intelligence training briefs to staff and replacement aviators.

In many other cases—applications for subspecialty, requests for entry into technical programs, and so on—a commander can cite specific experience directly related to the request. At other times, an endorser should try to connect the individual's general qualities with the particular request. Below, a ship's captain endorses the request of one of his JOs for a change of designator to Supply Corps. The captain points out aspects of the lieutenant's past performance that make him particularly apt for a supply career:

> A recent squadron command inspection rated his Admin Department OUTSTANDING, with Lieutenant Patel receiving special praise for his superb organizational skills, close attention to detail, and dedication to the concept of "service to the crew." All these qualities should serve him well in the Supply community.

Another possibility is to make specific comparisons to others. This technique is used well in the following endorsement to an officer's request for selection to civilian postgraduate school:

> Although Lieutenant Adams is ninth in seniority among thirteen talented lieutenants in this command, he is this squadron's number three lieutenant, a rating all the more remarkable since the two officers rated above him are board-eligible for O-4 this fall while Lieutenant Adams has only recently been promoted to O-3.

Then you might try other ways to state the degree of your support, because these statements can also help the selection board. Here, a commander endorses a chief's request to be considered for LDO:

> I would be pleased to have Chief Jones under my command as a commissioned officer, proud to have her in my wardroom, and gratified to know that the support establishment was in the hands of someone as capable as she.

Finally, let the board know the length and closeness of your observation of the person in question, if not already obvious. For instance, if you were recommending one of your Sailors for a special training program or an officer program, include something like this:

> I have known Petty Officer Anders for three years, two as her division officer and a third year working in a different division in the same department. Throughout this time, I have observed her remarkable professionalism, technical skill, and leadership.

Overall, the best endorsements are ones in which the facts themselves speak. The writer should provide strong evidence that the applicant possesses qualities appropriate to the opportunity requested. Specific, convincing details enable readers to see for themselves both the quality of the person being recommended and the fit between the person and the qualification being sought.

Responding to a Personal Request

Throughout the services, officials of various ranks must compose replies to personal letters from service members. In the Navy the heaviest load of this kind falls on BUPERS (which must reply to lots of letters from parents, too). But many other offices must respond to service members who have made personal requests.

Being considerate, on the one hand, and being fair, on the other—at the same time remembering the "needs of the Navy"—is obviously no easy job. In a memo at BUPERS, ADM David Harlow once specially outlined the need for written communication in this situation:

> Too frequently, correspondence crossing my desk indicates that in a telephone conversation, someone within BUPERS has reportedly made a commitment to an individual who has become difficult to support in terms of current policy. Sometimes, the cause may be a sympathetic desire to make a situation more acceptable to a constituent; sometimes, a misconception can arise when

---

FIGURE 2.10. *Letter on Behalf of a Personnel Issue*

From: LT James Helpful, SC, USNR

To: Whom It May Concern

Subj: ENTITLEMENT TO EDUCATIONAL BENEFITS

1. Formerly the Disbursing Officer for NMCB-1, I am writing to express my concern that LT Susan B. Frustrated may be denied her Montgomery GI Bill benefits as well as $2,400.00 of her pay due to a number of disbursing difficulties beyond her or any service member's ability to correct.

2. Dutifully aware of the requirement to pay $2,700.00 to convert from VEAP to Montgomery GI Bill (MGIB) benefits, LT Frustrated completed an allotment form in May 2001 in my office to begin monthly deductions from her base pay starting in June 2001. Unfortunately, our computing systems rejected my Disbursing Clerk's numerous attempts to process the allotment.

3. I assigned LT Frustrated's allotment exclusively to one of my most diligent disbursing clerks to resolve, and with her almost daily attention, the allotment took nearly three months to begin. Apparently (in the least disbursing-specific terms I can manage), the newness of the conversion program prevented the allotment from being recognized when it was released to the Defense Finance and Accounting Service (DFAS). Rigid bimonthly cut-off dates, unfulfilled promises of weekly resolutions from DFAS technicians, and operational commitments such as field exercises that removed my staff from the office delayed the start of her MGIB allotment until August of the same year.

4. Ultimately, the two- to three-month delay would prevent the $2,700.00 required for LT Frustrated's conversion benefits from being paid by October 2002, a deadline of which both she and I were unaware.

5. Finally, though an emotional plea has little place in a letter of this nature, I would kindly ask you to review LT Frustrated's MGIB account favorably, for I am beholden to her officership like few others in the U.S. Navy. I had already completed my undergraduate and graduate education when I arrived at my first duty station, where LT Frustrated served as the Admin officer. Her guidance to me and every other junior officer in our battalion was inestimable. I can only imagine what she could do for the rest of the fleet with benefits afforded by higher education through the MGIB.

6. Thank you for your consideration in this matter. If you require any additional information, please contact me at (410) 293-6225 DSN 281-6225 or U.S. Naval Academy, English Department, 107 Maryland Avenue, Annapolis, MD 21402-5044.

Very Respectfully,

J. Helpful

the individual "hears what he/she wants to hear." When we try to reconstruct what was meant as well as said, there is often no record. It has happened to all of us. In any case, our credibility, if not integrity, can be severely damaged.

ADM Harlow suggested that often the best answer to questions asked by phone is to say that you will research the problem and respond in writing.

Often, a well-written letter can make a tremendous difference. Along a somewhat different line, a naval officer once asked the disbursing officer of her unit to write a generalized "To whom it may concern" letter concerning her frustration in receiving the Montgomery GI Bill benefits to which she was entitled. This difficulty had not been resolved upon the disbursing officer's reassignment. In that letter (adapted in figure 2.10), the disbursing officer outlines the problem clearly.

Letter to a Promotion Board

As noted above, MyNavyHR.mil offers some templates for letters to a promotion board. But you must do your best to write with the particular communication situation in mind. Guidance regarding Navy officer boards stipulates that any such correspondence must contain a written acknowledgment by the eligible officer that he or she desires it included in the record; the rest of what to include is up to you.

What governs what to say and the impression your letter will make? Returning to the triangulation principles of chapter 1 for any writing situation, consider your audience, your purpose in writing, and who you are with respect to both:

Audience. Remember, a promotion board is composed of individuals senior to those under consideration. These individuals have wide naval experience, so you don't need to explain to them how the service works. On the other hand, they will not all be experts in your particular career field. On a Navy chief's board, for instance, some rates will be unrepresented, and so a few technical details may have to be explained. But always remember that board members typically have a great deal to do in a short time. See chapter 5 for more details about how writing evaluations relates to selection boards.

Purpose. What is your message? This is *not* simply that you should be promoted. That desire on your part is understood, and the decision is the board's. Your goal in writing to the board is *to explain circumstances in your record that do not stand on their own.* In fact, you shouldn't write at all unless there are some special circumstances or facts the board doesn't know or might overlook about your past performance. What kinds of circumstances might qualify? These would include unusual duty, for example, or an unusual pattern of career assignments. A logical explanation of broken service might be well received. Of course, there are many other possibilities.

Writer. Who are you with respect to the board and the purpose at hand? This may seem simple, but remember that you are under consideration and junior to your audience. You are a member of the Navy or Marine Corps and should sound like one. Write with humble respect in a confident, professional tone. You must not appear to be looking for any advantages but simply for fair treatment.

Below are a few DOs and DON'Ts for this kind of letter:

Do:

- Be as brief as the subject matter allows. (Long letters, regardless of content, are viewed dimly.) Write a page or a page and a half at most.
- Write respectfully to the board members. Assume they are scrupulously fair.
- Get to the point quickly.
- Spell out the vital facts explicitly and clearly, putting them in logical order.
- Explain unusual circumstances.
- Explain the positive aspects of any negative facts.
- Have your letter reviewed by former selection-board members, if you can, or other experts such as senior officers and leading chiefs.

Don't:

- Explain what board members will already know.
- Make grammatical or spelling errors.
- Express negative opinions or emotions either toward the naval service or toward particular individuals.
- Write a letter to the board at all if you have no clear reason to do so.

It is also possible for someone other than the service member to write the board on that individual's behalf. Figure 2.11 presents a letter written by a Marine Corps captain. A lieutenant who had once served with him (the names are changed) asked the captain to write to the promotion board on his behalf. The JO had received a poor fitness report from a Marine general during Desert Storm—unjustly, in his opinion. He worried lest he not be selected and then be dismissed from the Corps.

FIGURE 2.11. *Letter to a Promotion Board*

12 Jan 94

From: Captain Steven S. Alpha

To: President, 1994 Captain Selection Board

Via: First Lieutenant J. R. Bravo xxx-xx-xxx/xxxx/xxxx

Subj: INFORMATION FOR CONSIDERATION: CASE OF
FIRST LIEUTENANT JAMES R. BRAVO xxx-xx-xxxx/xxxx/xxxx

1. This letter is forwarded to the 1994 Captain Selection board for consideration in the promotion of Lieutenant Bravo. I was Lieutenant Bravo's company commander from 13 July 1992 to 1 November 1993 at Company x, xx Light Armored Infantry Battalion. I feel I can present an accurate picture of Lieutenant Bravo's value to the Marine Corps.

2. Lieutenant Bravo served as a scout platoon commander, weapons platoon commander, fire support commander, and executive officer of my company. He performed his duties with distinction and enthusiasm. Lieutenant Bravo exhibited exceptional knowledge of light armored reconnaissance employment and fire support at both the platoon and company levels— knowledge superior to that of his peers. Additionally, Lieutenant Bravo inspired unparalleled devotion from his Marines—a loyalty, in fact, greater than I have ever witnessed. As acting company commander on several occasions, Lieutenant Bravo displayed maturity in dealing with higher headquarters, forethought in integrating training with maintenance, and determination to succeed at every task. His performance was equal to the challenge of command and irrefutably showed his potential for promotion.

3. Lieutenant Bravo demonstrates the qualifications of leadership, knowledge, and professionalism. Based on my observations, his worth to the Marine Corps cannot be overstated. I enthusiastically endorse his advancement to the grade of captain. KEEP THIS MARINE OFFICER IN THE CORPS!

S. S. Alpha

Note that, administratively, you would need to submit this type of letter as an enclosure to your own letter to the board. The members will not consider it if your colleague or senior officer submits it directly.

Letter of Instruction

The personnel-related letter of instruction (LOI) is a nonpunitive warning to a service member about subpar performance. You can issue the LOI either to an enlisted person or to an officer. It amounts to formal counseling documented in writing. The LOI is a variation of a Navy Page Thirteen or a Marine Corps Page Eleven entry in a service record, but the letter is written in a more formal style than those because it is addressed directly to the individual.

Some commanders issue an LOI only to document performance poor enough to lead to the service member's relief or transfer to other duties. Others use it as a preventative measure and issue one as early as possible when a service member's performance first begins to degrade. Those who take the latter approach argue that issuing one at the eleventh hour is a way to get rid of a problem, not to solve it.

FIGURE 2.12. *Letter of Instruction*

1611
21 Dec 21

From:　Commanding Officer, USS VESSEL (DDG XYZ)
To:　　Ensign J. R. Officer, USN, XXX-XX-XXXX/1160
Via:　　(1) Executive Officer, USS VESSEL
　　　　(2) Weapons Officer, USS VESSEL

Subj:　LETTER OF INSTRUCTION TO CORRECT SUBSTANDARD PERFORMANCE

1. Your professional growth as a Surface Warfare Officer has been substandard. Specific deficiencies include:

    a. Lack of satisfactory progress toward surface warfare qualification.

    b. Inadequate knowledge of the 5"/62 gun system and its operation and maintenance requirements, though you attended Gunnery Officer School and Ammunition Administration School.

    c. Inadequate knowledge of the 3-M system, manifested by improper planned maintenance scheduling and performance.

    d. Inability to develop, plan, and execute a basic division training program despite specific direction from your department head.

2. To improve your performance and help you become competitive with your contemporaries, I direct you to complete the following actions by the dates specified:

    a. Learn the mechanics, operation, and capabilities of the 5"/62 MK 45 MOD 4 gun, and be prepared to demonstrate that knowledge no later than 31 Jan 22 at an oral board composed of LCDR Grit, LT Grapple, and LTJG Grog.

    b. Complete Basic 3-M PQS (NAVEDTRA 43241D1) prior to 31 Jan 22.

    c. Demonstrate significant progress toward Surface Warfare Officer qualification in the areas of division officer and warfare qualification. You will establish a qualification time table and submit it to the Senior Watch Officer, the Weapons Officer, and the Executive Officer no later than 5 Jan 22.

    d. Complete the Gunner's Mate Guns (GMG) Petty Officer 3 & 2 correspondence course no later than 28 Feb 22.

3. These requirements constitute the bare minimum necessary to improve your SWO knowledge and skills. Completing these requirements will also help you gain a more confident demeanor when dealing with juniors and seniors.

4. Individuals in your chain of command are always ready to assist you or provide additional guidance. Request help from me, from the Executive Officer, or from the Weapons Officer whenever you need it.

　　　　　　　　　　　　　　　　　　　　　　　　D. D. Skipper

Composing an LOI serves two major purposes. First, having to write one forces superiors to determine and describe the specific behaviors that need to be corrected. Second, having to read one requires the service member to direct attention to problem areas and to explicit methods of improvement. In the best case, the formality of such a written set of instructions, including well-defined goals and final attainment dates, can help galvanize a service member into great effort.

As with performance appraisals and awards (discussed in chapters 5 and 6), the cardinal rule in writing an LOI is to be specific. General comments that simply convey, "you're a poor performer and need to build up your character and to improve in discipline" just demean the individual and offer no real assistance. Detailing both the particulars of past deficiencies and the specifics of the needed improvement can make the LOI a highly effective counseling tool.

The LOI should outline these five matters:

1. The specific failures in performance: when, where, and what.

2. Actions the command has taken to correct this problem.

3. Steps the service member must take and any goals he or she must meet for each area of failing performance.

4. Individuals who can help.

5. Deadlines for each step.

If well thought out and accompanied by the right kind of personal counseling, an LOI can genuinely aid the service member, helping greatly improve her or his performance. On the other hand, if the service member refuses to be helped or for some other reason cannot meet the goals on the dates specified, the letter serves as specific documentation of the effort to help and the service member's continued failure. It then becomes formal grounds for subsequent relief, separation, transfer, disenrollment, or other such proceedings.

Figure 2.12 is an example of an LOI written to an officer who has gotten behind in his warfare qualifications, including division-officer administration.

Figure 2.13 is an example adapted from an actual LOI that assigns extra military instruction to a petty officer for generally poor performance. The XO who wrote the original version said this was a "low-level" LOI that, he hoped, would help round this service member into shape.

Notice in the second example its direct, unambiguous language. Although the letter does not berate the Sailor, it makes clear the shortcoming. In addition, notice the specificity of the instructions FT2 Smith must take, including the deadlines. You may find it difficult to write a direct letter, and it may feel critical, but this step is

18 Nov 20

From: Executive Officer, USS VESSEL (DDG XYZ)
To: FT2 (SS) Smith

Subj: LETTER OF INSTRUCTION TO REMEDIATE WATCHSTANDING DEFICIENCIES

1. You have been assigned Extra Military Instruction as a result of your poor watchstanding practices and your lack of attention to detail. Accordingly, you must complete the following actions:

   a. Study the duties of the Below Decks Watch (BDW) specified in the SSORM (Art. 2305), with particular attention to what the Below Decks Watch must and must not do. Due: 12 Dec

   b. Read the following selected articles from Naval Reactors Technical Bulletin (NRTB) Volume B, Revision 2, Book 1 (a copy of each article has been left in a folder with the Chief of the Watch for your convenience):

	Article	Page(s)
(1)	"Zero Defects"	B-1
(2)	"Be Alert"	B-3
(3)	"Violations of Proper Watchstanding Practices"	B-34, 35
(4)	"Bilge Alarms"	B-42

   c. Complete an oral interview with your Department Head in which you discuss your duties and obligations as the Below Decks Watch and the lessons to be learned from the reading assigned in paragraph (b). Due: 15 Dec

   d. Conduct a two-hour monitor watch of another BDW, as specified on the watch bill generated by the Chief of the Boat, with emphasis on procedural compliance, attentiveness to irregularities or potential problems, the watchstanding routine, familiarity with the watch station, and accuracy and interpretation of logged readings. Due: 18 Dec

2. This remedial program is designed to make you think about the responsibilities and watchstanding practices of the ship's BDW with the long-term goal of enhancing your BDW watchstanding performance as well as that of other BDWs who may be similarly deficient.

E. James

designed to help the Sailor make improvements before the next evaluation. Ideally, if Petty Officer Smith will follow these steps and show the kind of improvement desired, this shortcoming will not need to be formally documented in his or her evaluation. Alternatively, if he or she does *not* improve, the documentation in this LOI will serve to justify further, more serious documentation and disciplinary steps.

BUSINESS LETTERS

The business letter should really be called "the business or personal letter," since, in addition to business correspondence, you can also use it for sending thanks, congratulations, or condolence. As the *Correspondence Manual* points out, it can even

FIGURE 2.14. *Business Letter Guidance*

Department of the Navy
USS HALEAKALA (AE 25)
FPO SAN FRANCISCO 96667–3004

5216
Ser AE 25/28
January 10, 2022

Business or Company Name
Attn: Person within the Company
Street Address
City, State ZIP

THE FORMAT, STYLE, AND USE OF A BUSINESS LETTER

Bottom Line Up Front: State the purpose of your business letter in the first paragraph unless the occasion calls for delay to soften bad news or to introduce a controversial proposal.

Keep paragraphs to about 10 lines each, with the first being much shorter than that. Work at writing even more clearly in letters to civilian audiences than in standard naval letters. In particular, avoid all unfamiliar military terminology, including acronyms. If you must use such specialized language, make sure to explain each term the first time you use it. Many things that we take for granted in our internal correspondence is unfamiliar to an external audience.

A few other things differentiate a business letter from a standard naval letter:

1. Don't number main paragraphs, but you may number subparagraphs like this one for ease of reading or when making a list.

2. There is no "From" line on a naval business letter, so you must use a letterhead (printed, stamped, or typed) to show the letter's origin.

3. You can express dates in month-day-year order if your readers are likely to be most familiar with that format. Whatever the order, always spell out or abbreviate the month rather than designate it by number. (If you express the date as 3/4/22 or 11/12/21, some readers may mistake the day for the month or vice versa.)

4. If writing to a company in general but directing the letter to a particular person or office within it, use an attention line between the name and address. See above.

5. You may use either a salutation or a subject line in a business letter. If you use a salutation, be alert to the gender of your audience, and follow the *Correspondence Manual*'s specific guidance. On routine administrative matters, you may replace the salutation with a subject line (as above). Used this way, a subject line orients readers to the topic and avoids the stiltedness of "Dear Sir or Madam."

6. Business letters typically use the complimentary close "Sincerely," whereas standard naval letters omit the complimentary close entirely. If you wish to show special deference to a high public official, you may replace "Sincerely" with "Respectfully" or the like.

Sincerely,

LT O. H. Perry,
U.S. Navy Administrative Officer
By direction of the Commanding Officer

Encl:
Correspondence Manual (sep cover)

be used "for official correspondence between individuals within the Department of Defense when the occasion calls for a personal approach."

You can vary the format of a business letter somewhat so that, say, a letter of condolence isn't encumbered with serial numbers, or so the text of a short letter fits in the middle of the page. The main variations, however, result from the differing uses of a business letter. We'll discuss several cases beginning with what is probably the most common use.

Business Letter to a Business or Other Organization

Figure 2.14 provides guidance on how to write a business letter—in the form of a business letter.

Good-Will Letter

You will often want to express thanks to a person in or outside the military who has done you a favor. You might be thanking someone for a talk, for a personal favor of some kind (such as an introduction), for hospitality, or for any of a number of other services.

Whatever the circumstances, be genuine. Avoid form letters in this kind of writing, and strive especially to express real gratitude. The latter takes some care. Sometimes informality will help you write genuinely, with the degree of informality depending on your relationship with the correspondent and the specific situation. Recount some of the details of the service rendered and the specific results of the person's favor, if possible.

Make sure your letter doesn't appear stuffy or insincere, as if you are just trying to conform to the rules of service etiquette. To this end, avoid the passive voice. Saying "it was appreciated" instead of "we appreciated" will communicate aloofness,

---

FIGURE 2.15. *Good-Will Letter to Local Business*

Dear Mr. Narita:

On behalf of the men and women of the Personnel Support Activity Detachment, I want to thank you for your gracious hospitality on Saturday, October 10.

From the moment you met us at the train station until our departure from your lovely restaurant, we knew we were in excellent hands. We will never forget the warmth of your reception. As a result of this trip, we are all eager to visit more places in your beautiful country.

Please extend our gratitude to your family, especially your son, who so willingly took us to our destination. The day was truly blessed with good weather, good food, and good memories.

Sincerely,
Captain Alice Hayward
United States Navy

not gratitude. Of course, be timely in thanking someone; write quickly so you won't forget about it—at least within forty-eight hours of the occasion.

Figure 2.15 comprises the text of a letter of thanks written to a Japanese restaurant owner by the officer in charge of a small Navy unit. The writer speaks her gratitude warmly and simply.

Letter of Appreciation

Within the Navy, a letter expressing thanks for a job well done can provide encouragement and praise to someone who has supported your mission but does not necessarily work directly for you. Figure 2.16 is a letter (fictitious but adapted from a real letter) to someone in a private business, expressing thanks to the representative

---

FIGURE 2.16. *Letter of Appreciation*

DEPARTMENT OF THE NAVY COMMANDER
NAVAL METEOROLOGY AND OCEANOGRAPHY COMMAND
1100 BALCH BOULEVARD
STENNIS SPACE CENTER MS 39529–5005

> 1650
> Ser 0/213
> 29 Nov 2019

From: Commander, Naval Meteorology and Oceanography Command
To: Mr. Mike Abed
Via: Mr. Lee Stevens, Environmental Systems Research Institute, Inc.

Subj: LETTER OF APPRECIATION

1. Please extend my sincere appreciation to Mr. Mike Abed for his contribution during our recent Fleet Battle Experiment (FBE-I).

2. Faced with the Oceanographer of the Navy's challenge to integrate Meteorology and Oceanography (METOC) products in operator systems, the Naval Pacific Meteorology and Oceanography Command, San Diego, California, needed to rapidly integrate METOC information into the Naval Fires Network (NFN). NFN is a new but critical system. Completing the task required accomplishing twelve months of work in four months.

3. Mr. Abed's team allowed us to satisfy 100 percent of our short-fused customer requirements, an accomplishment previously thought to be impossible. As a result, Commander Third Fleet has asked us to expand our efforts, in addition to delivering this level of support on a continuous basis.

4. Mr. Abed was, without question, one of the key reasons for our unprecedented success. He is an outstanding individual, an enthusiastic professional, and a true asset to any organization. He is welcome here any time!

5. You have my sincerest thanks.

> B. Donaldson
> Rear Admiral, U.S. Navy

of a company who has done important work for the Naval Meteorology and Ocean-ography Command. Notice that while this letter is addressed to the individual, Mr. Abed, the text actually addresses his supervisor, through whom it is routed (Via:), drawing Mr. Stevens' attention to his good work.

With a ghostwritten letter of this kind, if the flag writer has done her or his job well, all the admiral has to do is to sign and perhaps pen in a personal note.

Letter to Parents and Other Family Members

Parents of Sailors and Marines are a type of audience different from those within the military. Supportive of their children, usually patriotic, and also understanding of the scrapes their children get into, parents will usually be responsive if com-manders treat them well. A Navy commander who reviewed an earlier edition of this guidebook commented about his own efforts in this area:

> Writing letters to family members helps more than you can imagine. When I was OIC [officer in charge] of Naval Security Group Detachment Diego Gar-cia, I wrote letters to parents, spouses, adult children, or whoever was listed as next of kin in the service record. Promotions, awards, and when the member first checked into the command were the occasions that generated my letters.
>
> Many of the parents wrote me back. One mother demanded to know why there had never been an article in the local paper about her daughter's many promotions, awards, and achievements in the Navy. A former Navy dad congratulated me on being an LDO (although I'm not) since no "real" officer would have the leadership to write letters home. Many Sailors thanked me personally for this minimal effort. Many more thanked my assistant OIC.
>
> One Sailor told me that his parents were so proud that they framed the letters and hung them on the wall next to his high school graduation picture. Sailors felt more connected to the rest of world. It is hard to prove these things empirically, but I believe we had significantly fewer discipline problems than other detachments/commands of similar size because of this letter-writing effort.

Write to parents and other relatives of your people often. Welcome them to the unit's family. Praise their young service members on their promotions and awards. Always assume in your writing that your audience is mature, intelligent, and understanding.

These letters will help keep up morale, and it's simply a good thing to do. It will begin to establish a relationship. On more difficult subjects that might come up later, it's always easier to write to a familiar audience, one with whom you have already established contact.

One good example is provided in figure 2.17, adapted from a letter in which a CO congratulates the parents of a petty officer on her recent promotion to second class.

Another example, provided in figure 2.18, addresses the children of a service member who has been doing excellent work while deployed. Notice how the admiral

---

FIGURE 2.17. *Letter to Parents*

[COMMANDING OFFICER LETTERHEAD]

December 18, 2022

Dear Mr. and Mrs. Loh:

You may have heard the good news already from your daughter Sami, but I would like to officially inform you of her recent advancement to Boatswain Mate Second Class (E-5). She has demonstrated exceptional initiative and personal effort in reaching this goal on the advancement ladder.

Her appointment as a Second Class Petty Officer carries with it the obligation of exercising increased responsibility and new roles. I have every confidence that she will discharge the duties of her new position with the same dedication to duty she has displayed in the past.

Sami is an outstanding Sailor who has set an excellent example for the junior personnel aboard. The Navy needs more women and men of her caliber. I am sure you are as proud of Sami's achievement as we are. Best wishes to you.

Sincerely,
K. V. Lumen
Commander, U.S. Navy

FIGURE 2.18. *Letter to Family Members about a Service Member's Good Work*

August 5, 2020

Dear Alan, Joseph, Abby, and Alice,

I want you to know how proud we are of your father, who works for us here in the Bureau of Naval Personnel. Not only does he do an excellent job every day, but he also spent a lot of his own time, after work, putting together and watching over a group of 26 people during the trouble we had in Guantanamo, Cuba, last month. Because of your father's hard work and careful planning, somebody from that group was always here to answer questions and solve problems for our Navy families in Cuba as they all got ready to leave. That wasn't an easy job, but your dad rolled up his sleeves and made sure the job was done right.

Your dad did such a good job that I gave him a special award (called the Navy Achievement Medal) during a big ceremony here at the Bureau. I wish you could have been there.

Because of your dad's service in the Navy, that makes you special—but then, you always were and always will be special. I hope you'll come to visit the Bureau sometime so you can see where your dad works. It would be great to meet you.

Your friend,
R. J. Murphy
Vice Admiral, U.S. Navy

Master Alan L. Sailor
Master Joseph W. Sailor
Miss Abby B. Sailor
Miss Alice A. Sailor
8222 Housing Place
Manassas, VA 22110

writing the letter has adopted a tone and style appropriate to his audience. We can be sure most of his memos or directives to those under his command were worded quite differently.

Letter of Condolence

> *"Few will go through a career without the loss of a person in a unit—*
> *a friend, a shipmate, or a subordinate. The parents, spouse, and children may*
> *have very little to comfort them; sometimes a sincere letter that expresses what*
> *the loved one meant to another will be treasured. But expressions of insincerity*
> *may be worse than not being heard from at all."*
>
> —NAVY CAPTAIN

If a service member has died and you must write to express your sympathy to a spouse, to parents, or to children, how do you go about it? This is no occasion for formulas—no form letters, no canned phrases. Don't copy an example from a book, including this one. As the captain quoted here notes, anything that sounds insincere will be worse than writing nothing at all.

Express sympathy, sadness, or compassion. Say what you can positively about the dead; say what you can in the way of sympathy with the living. You might mention the loss that shipmates feel. Perhaps the best guide is to search your own feelings and to remember your experiences with the person who has died. Reflecting on something you've shared with their deceased loved one or something you know of the family might provide you an appropriate subject.

Be brief. Don't philosophize on the meaning of life and death; quote other texts sparingly—and only if you know your particular audience will receive it well. Service to country or shipmates might be proper topics, or perhaps the service member's cheerfulness, dedication, or good deeds. Use your best judgment. You might express your willingness to do whatever you can to help. Usually, your own sympathy for the bereaved, your shared knowledge of their loved one, and your understanding of their pain are the chief expressions that might be of comfort. Figure 2.19 is the text of an actual letter of condolence written by LCDR Dan Gallery to an admiral friend. Although written nearly a century ago, it remains a model of compassion and authenticity on a difficult subject.

Keep your tone familiar. Speak in the first person and use first names where appropriate. Show the letter to others prior to sending it, if you have any doubts, to see how what you have written strikes them. Sometimes chaplains can help with the writing. Pay particular care to the preparation of the letter, whether you type

---

FIGURE 2.19. *Letter of Condolence*

January 4, 1935

Admiral Joseph M. Reeves, U.S.N.
USS Pennsylvania
San Pedro, California

Dear Admiral Reeves:

Mrs. Gallery and I send our deepest sympathy to you and Mrs. Reeves in the loss of your splendid boy. As the years go by, our chief interest is in our children. They make life worth living. They give us a new pleasure our earlier years never knew. We glory more in their success than in any that may have come to us. You have attained the highest place in the service, a place few indeed reach, but the promise of your boy gave you just pride and joy beyond any brought to you by the rewards gained by your brains and industry.

We understand what grief is yours. But know that we and your wide circle of friends grieve with you.

Yours truly,

Daniel V. Gallery, Jr.

---

it or write in longhand. Keep the format simple; don't include a serial number or otherwise clutter it with bureaucratese.

Besides your personal letter, there will probably be an official one from the command discussing funeral arrangements, shipping of personal effects, a command memorial service, and so on. The person who writes that letter will have to pay scrupulous attention to the accuracy of all the details and should express sympathy too. But as a commander, leader, or friend, take the extra step of writing a personal letter. Don't mix up your genuine condolence with mere officialdom.

Letter to a New Command

Writing a letter to a new command introducing yourself to your new CO before you arrive can make a positive, professional first impression. Such a letter is not just a courtesy but serves several distinct purposes. To start with, it alerts a ship or other station to the pending arrival of someone's relief. Normally, the command will have already heard through the personnel system of an assignment, but even if they know your name, they usually won't have much information on your specific qualifications, background, or interests. Nor will they necessarily know anything about your family status, your need for knowledge of the area (for housing, schools, and so forth), or your plans for leave and arrival dates.

You will likely make informal contact with your new command over email as well, but the thoughtfulness and personal touch of a more formal letter (even if sent electronically) can make a difference. In addition, a letter is a bit more proper than an email and is a bit less likely to be overlooked in the dozens of messages that a CO typically receives each day.

If you can fill in your future seniors and shipmates on this kind of information, it may help them fit you into billets, ensure for a contact relief, arrange an appropriate sponsor to help you find housing, and so on. They might be able to schedule various kinds of helpful temporary duty if you give them enough information. Moreover, penning such a letter can create positive first impressions with both officers and enlisted. Not long ago, a Navy lieutenant with orders to CINCLANTFLT staff sent

---

FIGURE 2.20. *Letter to a New Command*

> 1265 Kentucky St.
> Lawrence, Kansas 66044
> 913-xxx-xxxx
> cquinn@emailprovider.com
> 22 June 2022

Commander Robert Jones
USN Commanding Officer
USS Nitze (DDG 94) FPO xxxxx

Dear Commander Jones:

I recently received orders to the USS Nitze. I am delighted to be assigned to a fine destroyer like yours and greatly look forward to reporting on board.

I will graduate from the University of Kansas in late July and am scheduled to report to Surface Warfare Officers School the first of September. While at Kansas, I have been rush chairman, vice president, and president of my fraternity chapter (Phi Gamma Delta), as well as captain of the fraternity's flag football team. I also sang in the University Concert Chorale, captained a quiz bowl team, and am a member of Sachem Honor Society. As you probably know, my major is history, and my minor is English. In the NROTC unit here at Kansas, I was company officer my senior year; I had summer training on the USS John C. Stennis (CVN 74) and USS Lassen (DDG 82).

Just after graduation, I will be getting married to Jeannette Ralph in Liberty, Missouri. Jeannette graduated from the University of Kansas in May and has done a good deal of work as a technical writer. We enjoy swimming, scuba diving, and biking.

I realize that the billet to which you assign me must be based primarily on need. Yet given a choice, my preference would be a first billet in either Operations or Weapons.

Thank you for the information you have sent me. Unfortunately, it does not appear that I will be able to take advantage of your offer to visit the Nitze until after reporting to SWOS. If my plans change, I will let you know.

You can reach me at the above email address at any time till my reporting date. Also, I will continue to receive mail at the Lawrence address until leaving for SWOS in late August.

> Very Respectfully,
> Charles Quinn
> Midshipman First Class, U.S. Naval Reserve

a short letter to the admiral, well-crafted and professional in appearance. The letter impressed the deputy, who sent it on to his boss. The admiral responded, "I want to meet this lieutenant when he comes in. Give this officer an arrival call." The lieutenant's letter had set him on a fast track.

The example in figure 2.20 is a fictional letter from a Navy ROTC graduate to his first command.

Letter of Recommendation or Reference

Commonly, a commander, division officer, or chief will have to write a letter of recommendation to a corporation or school outside of the Navy. Someone may

---

FIGURE 2.21. *Letter of Recommendation*

22 Jan 2021

University of Florida
Civil and Coastal Engineering Dept.
124 Yon Hall
Gainesville, FL 32611

Dear Sir/Madam,

LTJG Charles Engineer carries my strongest personal recommendation for admission into your graduate school program. I have been his Commanding Officer for nearly two years, during which time he consistently exceeded my expectations.

LTJG Engineer is an intelligent and highly motivated Naval Officer who displays exceptional maturity and an inherent ability to lead. He gained my utmost confidence during his independent position as Officer in Charge of 26 enlisted personnel during a seven-month deployment to an isolated island over 1,500 miles away from my location. While deployed, he was engaged in two major construction projects that provided essential housing and quality of life improvements for the local population.

I recognized Charles early as a "quick study" and someone who easily retains large amounts of information relevant to his assigned duties. He completes all tasks on time and has uncanny attention to detail. Charles's ability to effectively prioritize large workloads and routinely produce creative solutions to complex problems suggest to me that he would be a proactive, competent student in a research-based program.

His Bachelor of Science in Ocean Engineering and his ability to work through tough, stressful situations provide a firm foundation capable of completing the most demanding Graduate School curriculum. His intense enthusiasm for ocean-related study is evident in his recent successful application into the Navy's Ocean Facilities Program. When he completes graduate school, Charles Engineer will be one of only a few Ocean Engineers providing valuable expertise to the Navy on countless coastal and ocean-related facilities.

Finally, as a Florida alumnus, I can assure you that he will continue to represent our fine institution with distinction.

Sincerely,
A. J. Architect
CDR, CEC, USN

have retired or left the service and asked for a letter. Also, several naval programs exist that pay for an individual to attend a civilian college or training program that prepares them for further naval service. To gain entry into a civilian school, that person will have to ask for several letters of recommendation.

There's no mystery about writing letters of recommendation, but it can feel intimidating at first. A writer should usually speak to the qualities being looked for in a specific program or profession. And in speaking to your knowledge of the individual, you may also want to mention the significance of your own credentials. In general, remember to translate military details into civilian terms, pointing out the importance of accomplishments or duties that might not be clear to someone lacking naval experience.

The letter in figure 2.21 was adapted from an actual letter written for a naval officer applying for graduate study in engineering at the University of Florida. The reference does well in making clear why the young officer is likely to succeed at graduate work, although much of what is said also pertains to the officer's likely success in his future naval career (also important).

The fact that the commander signing the letter is an alumnus of the university to which he is writing is a helpful touch, but it is not required.

CONGRESSIONAL RESPONSE LETTERS

A special kind of letter that virtually all commands must write sooner or later is the response to a congressional inquiry. Being able to compose such a letter quickly and fluently is obviously an important skill—and not just for senior officers. Commands of all sizes can receive such letters, and they must respond in short order. While the final letter is usually honed by the XO or CO, a senior enlisted person or a JO could be assigned the first draft.

Context: The Typical Situation

Most congressional inquiries to commands outside Washington, D.C., have to do with personnel. Ship's captains or unit commanders will not have to defend large issues of naval policy to members of Congress. Instead, they will normally have to handle inquiries about individuals' complaints.

Perhaps a Sailor has been turned down for the Navy diving program and has written a letter of complaint to his local congressman about that refusal. A petty officer may have complained to her representative about not being able to strike for a particular rating. Perhaps a Marine has written his senator, contending that he hasn't received the correct pay.

In such cases a congressional staffer will send a letter of inquiry either directly to the service member's command or to the Navy Office of Legislative Affairs (OLA) or HQMC, which will then write to the command. (BUPERS alone gets many hundreds of congressional inquiries annually.) Sometimes the complaint originally sent to the member of Congress is vague, emotional, and clouded in perception. Indeed, the service member may not even have been the prime mover in crafting the letter. A spouse or parent may have convinced the Sailor to write and may even have ghostwritten the letter.

Whatever the circumstances, the senator or representative normally is simply asking for information with which to respond to the complaint. True, such correspondence can have great visibility. But typically, the Washington official simply needs some perspective as to the problem, and the ship or station need not be on the defensive.

In nine out of ten cases, you simply need to provide the requested information and the command's perspective to give the Congress member what he or she needs for a reply.

Audience: Consider the Congress Member's Point of View

The member of Congress will, of course, want to do whatever he or she can to address a service member's valid concerns. How can you aid the legislator in that task? Quite often, whether a complaint is justified or not, all the means for redressing that complaint within the service have not been exhausted. You can be helpful by explaining to the Congress member exactly what the service member's options are, including the individual's next move. Moreover, you might be able to provide a point of contact (with phone number) or send along forms the service member may need to pursue a further option, thus giving the member of Congress specific aid in responding helpfully to the original complaint.

Below are a few simple guidelines to bear in mind when responding to a congressional (or White House) inquiry.

Respond Quickly

After receiving a letter, the Congress member's staff has to write to the naval command. When it gets the letter, the naval command must investigate and write back to the Congress member. Finally, the legislator's staff must write back to the constituent. (If there is an intermediate step via a liaison office in Washington, D.C., then even more time is lost.) You can see that months could easily elapse if each stage isn't executed quickly. The rule is for a command to respond within a specific number of workdays (five is typical) of receipt of the inquiry. If you believe you will need more time, respond with an "interim" reply within forty-eight hours to let the Congress member know. Many commands get most of their "final" replies out within a day of receipt.

Be Factual

Research the facts and then lay out in order the actions that both sides have taken, as well as the rationale behind the command's actions. If the problem is a long-standing one, a history of actions on the ship's or unit's part may exist, and you can catalog those events. Often, the XO and CO have been involved with the issue before and will have lots of information on hand. Maybe the division officer has recorded some information in an official notebook, or perhaps someone has written a memo for record about a particular Sailor's request or complaint or about a key counseling session. The files may contain past inquiries on the same or similar topics.

Whatever the situation, be accurate, research and re-research the facts, and be sure of what you are writing.

FIGURE 2.22. *Excerpt from Letter of Recommendation for a Security Clearance*

I have known LT George Appealing since I returned from deployment in January 2003. As squadron mates we have interacted both professionally and off duty on a daily basis since then. In addition to evaluating his tactical performance while I was a squadron tactical instructor, I have also observed his general professionalism as a Naval Officer and Aviator.

LT Appealing is dedicated to his profession and has achieved every qualification commensurate with his time in the Navy. Despite the uncertainty and frustration created by the ongoing appeal process, he continues to show this dedication as Division Officer for Detachment Two. I can confirm that he is a man of integrity.

I have observed his dedication to his family at numerous times both on and off duty. This is not irrelevant. Loyalty to family is a character trait that extends to other areas of life.

Reference (a) Sec. C.2 states that the standard for access to classified information or assignment to sensitive duties is "the person's loyalty, reliability, and trust worthiness are such that entrusting the person with classified information or assigning the person to sensitive duties is clearly consistent with the interests of national security, [and] there is no reasonable basis for doubting the person's loyalty to the Government of the United States." In my estimation, there is no reasonable basis for doubting LT Appealing's loyalty.

For your information, I am currently assigned to Helicopter Antisubmarine Squadron Light Four Six as Detachment Seven Officer-in-Charge. I have over 1,956 hours of flight time in five different aircraft types. At HSL-48, I completed two deployments as well as serving as acting Squadron Weapons and Tactics Instructor, Safety Officer, and Quality Assurance Officer.

R. A. Ducent
LCDR, USN

Explain the Command's Perspective

A Sailor may have written out of concern—or demand—for his or her personal rights. The command certainly respects those rights, but it is also the custodian of the service's rights, which amount to the Sailor's obligations. Usually, those obligations will not have been spelled out in the original letter. The Congress member will want to see the command's perspective on all sides of the issue.

To show that perspective in full, you may need to explain some key regulations or procedures and to outline exactly what violations of procedures may have occurred. Remember to delineate the service-specific information needed to understand this particular problem.

Start with the facts and proceed to explain the rationale for the command's actions. As you write, account as best you can for the service member's point of view and feelings. Admit forthrightly any mistakes that the command has made. Being straightforward, evenhanded, unemotional, and factual will help keep the tone right and help avoid defensiveness and inappropriate criticism.

Write in "Civilian"

Be careful not to confuse your reader by using naval jargon or acronyms. Switch gears, step back a bit, and address your reader as an intelligent civilian. Explain any acronyms you must use, but try to do without them in the first place. Use civilian-style dates (month-day-year) in the heading and throughout the letter. But mainly, try to write as if you are talking face-to-face with the individual to whom you're writing.

Get the Details Right

A few specific details may seem small, but overlooking them can undermine your credibility and the success of the letter:

- Be sure to address the Congress member correctly. Consult the *Correspondence Manual* or current guidance for the proper address for senators, representatives, and other senior officials.
- See appendix B of the *Correspondence Manual* for a comprehensive description of inside addresses and salutations.
- Normally, begin a letter to a member of Congress with a thank-you phrase, such as "Thank you for your letter of February 21 concerning Petty Officer Johnson's pay problem." This first paragraph serves the same function as subject and reference lines in a naval letter. By stating here the subject the letter will discuss, the date of the Congress member's letter, and the constituent's name, you help the congressional staff find the right file.

- Write in business-letter style. (Some commands prefer you use a comma rather than a colon after the salutation.)

- Enclose an additional courtesy copy along with the original when you send it to the member of Congress.

- Navy commands replying directly to a Congress member are required to send a blind copy of the final reply and copies of all substantive interim replies to the OLA in Washington, D.C., and to BUPERS.

Ask the Right People to Review

Once you have written your response, have it reviewed by your command's writing experts. Occasionally, the command may need to consult a naval lawyer, officers up the chain of command, or the respective office that handles legislative liaison in Washington, D.C.

In the Marine Corps, normally, a command's Office of the Inspector will prepare the smooth draft from the information it is given. In addition, almost all responses are sent via the Congressional Inquiry Section of HQMC, rather than going directly to the member of Congress from the command. Whatever the process, the responsible officer should ask to see a copy of the smooth draft before it leaves the command. That way he or she can make sure no errors have been made and can check to make sure the inspector's drafter has not taken too much license with the facts provided.

Example: Letter to a Congressional Representative

Figure 2.23 is a fictionalized letter to a congresswoman. The lawmaker had inquired about a serviceman's loss of leave and asked why he had not been allowed to reenlist despite his receiving an honorable discharge. The person upon whose behalf Congresswoman Capitol has inquired is addressed as "Mr." in the letter since he is no longer in the Navy. (Note that current policies or regulations concerning illegal drug use may differ from those outlined in the example.)

The letter in figure 2.23 demonstrates a few additional important elements of an effective congressional response:

- It claims strong evidence supporting the reasons for the "not recommended" reenlistment code and explains the apparent discrepancies clearly (honorable discharge but not recommended to reenlist; seemingly unjust loss of leave).

- It also sketches the main details of the chronology without going into every detail.

FIGURE 2.23. *Congressional Response Letter*

[LETTERHEAD]

March 15, 20xx

The Honorable Dorothy Capitol
House of Representatives
Washington, DC 20515

Dear Mrs. Capitol:

Thank you for your letter of March 10, 20xx, concerning Mr. Charles Sailor's discharge from the Navy due to marijuana use.

Some details of Mr. Sailor's case may help clarify the circumstances of his discharge. The urine sample he submitted on June 18, 20xx, tested positive for tetrahydrocannabinol (THC). The Navy Drug Testing Laboratory used radioimmunoassay for the initial sample screening and gas chromatography for confirmation. These tests yield results in which we have high confidence with no false positives in over three years of quality-control testing.

After initially exercising his right for court-martial, Mr. Sailor changed his mind and requested his case be heard at a nonjudicial hearing under Article 15 of the Uniform Code of Military Justice. I found that he had committed the offense of illegal drug use with which he was charged, and I awarded punishment. Mr. Sailor did not exercise his right to appeal this punishment.

Care for due process and granting Mr. Sailor's requests, including a polygraph examination that he terminated prematurely, resulted in an adjudication period of July 1 to October 13, 20xx. This period is longer than usual but is still within the 120 days required by the Manual for Courts-Martial. During this period. I did not allow Mr. Sailor to take leave because he was in a disciplinary status. (My command policy is not to grant leave to individuals awaiting disciplinary action, placed on restriction, or serving extra duty unless an emergency or hardship is involved.) As required by statute, Mr. Sailor lost all accrued leave in excess of 60 days at the beginning of the fiscal year.

We retested Mr. Sailor for drug use two days prior to his October 22 discharge, and his urine sample again tested positive for THC. We did not receive the results until after his discharge and therefore took no further action. Although he received an honorable discharge, we assigned a RE-4 (not recommended) reenlistment code because of drug use, in accordance with Navy policy.

Mr. Sailor has the right to petition the Board for the Correction of Naval Records (BCNR) regarding his reenlistment code and loss of accrued leave. I am enclosing the necessary forms should he desire to do so. We established the BCNR for the purpose of reviewing naval records and correcting possible injustices.

If I may be of further assistance, please let me know.

Sincerely,
L. N. Officer
Commanding Officer

Enclosure

- The writer adds credibility by citing "three years of quality-control testing" (which suggests that the Navy is attuned to the possibilities of injustice) and by citing Navy rules (the assignment of a "not recommended" enlistment code; the statute on loss of accrued leave).
- The discussion of rights and of the ex-serviceman's logical next step manifests the command's concern for the individual's rights. Including the appropriate forms is also helpful.
- Overall, the letter is to the point, factual, and relatively brief.

Additional Congressional Response Examples

Following are some adapted, anonymous excerpts from similar letters, illustrating various tacks you may want to take in a particular case.

Explain the History of Events

We approved Petty Officer Garcia's request to work in Public Works at Naval Station Norfolk in the hope that he could receive better training and be more productive in a large, maintenance-oriented organization. But he was apprehended in an attempted theft of government gasoline, and for this action, he was subsequently awarded punishment at nonjudicial punishment ("Captain's Mast"). Because of this incident and his generally poor professional reputation, the Naval Station Public Works Officer disapproved his temporary transfer.

Petty Officer Smith's request for humanitarian reassignment certainly deserves consideration; however, until I received your letter of March 6, 2020, neither I nor anyone else in my chain of command was aware of his desires. He had not discussed humanitarian reassignment with his superiors, nor had he forwarded any written request for consideration. Petty Officer Smith has since stated that he wrote to you after discussing this issue with another shipmate (who is not in his chain of command) because his shipmate didn't feel that the request would be favorably endorsed.

Explain How a Specific Policy Applies to the Individual's Situation

The 4th Marine Division's policy gives commanders authority to reduce Marines in grade administratively because of unexcused absences. The unit commander sent to Lance Corporal Fairfax via certified mail a letter of intent to reduce him in grade. That letter informed the service member that he had 20 days to respond to his officer in charge about the unexcused absence allegations, but he made no attempt in that period to appeal the reduction. Lance Corporal Fairfax was then reduced to his present grade.

Write to Explain What Options the Service Member Has Not Yet Pursued

Petty Officer Threefoot is not eligible for normal reassignment until March 2024. Until then, he may consider a self-negotiated exchange of duty with another Sailor of identical paygrade and specialty who would like an assignment to Alaska. We have the necessary information if he is interested in pursuing this option.

Write to Explain the Command's Perspective

I wish to reiterate that we have made a concerted attempt to train Petty Officer Quartersell to perform satisfactorily at the petty-officer first-class level. But he must be prepared to dedicate a large amount of effort to self-study and to learning the basic tenets of leadership if he is to pursue a successful naval career.

Write to Set the Facts Straight

We process enlisted performance evaluations through the division officer, department head, and executive officer; then, they are signed by the commanding officer. Each level of leadership thoroughly examined Petty Officer Askew's performance record as did a review board of chief petty officers. He was, in fact, ranked as the worst petty officer second class in this command, and his performance was judged unsatisfactory. Despite Petty Officer Askew's assertions to the contrary, he is the only Seabee to receive an unsatisfactory performance evaluation during my 20 months of command.

Checklist for Preparing Answers to Congressional Inquiries

The Substance

- ✓ Does the letter get to the point, stating the basic response right after the thank you sentence?
- ✓ If the letter cites a chronology, does that chronology run smoothly and completely? Is the explanation sharp and pointed rather than rambling?

✓ Is the letter fair to the service member, and does it also appear to be fair? Does it avoid a defensive tone?

✓ Is the letter fair to the needs of the service?

✓ If mistakes have been made, have you stated them forthrightly and apologized or stated future compensatory action, as appropriate?

✓ Does the letter raise any issues it doesn't have to raise?

✓ Does the letter obligate the service to do something? If so, have you checked to make sure that the service both can and will do it?

✓ Have you given the member of Congress sufficient perspective?

✓ Does the letter answer all the Congress member's inquiries?

Technical Details

✓ Have you written this response in business-letter format, omitting the originator's code and the letter serial?

✓ Have you double-checked the addressee and address?

✓ Did you use "The Honorable"? Have you addressed a committee chair properly?

✓ Is the ZIP code correct? Is the salutation correct?

✓ Is the letter's opening stated properly?

✓ Have you included extra copies, as required?

✓ Have you checked for unnecessary jargon or acronyms?

3

Staff Writing and Operational Documents

"Lots of staffers produce packages that look terrific and are formatted perfectly but don't answer the original question."

—NAVY CAPTAIN, OPNAV

EXECUTIVE SUMMARY

If you work on the staff of a senior officer, you will likely be drafting documents specifically for the signature of your boss (or higher) or for other use by that senior officer. A number of specific document types support the administrative and operational functions of the Navy, and others play a key role in decision making. These often take the form of specific types of memos or directives. This chapter covers some of the more common documents you may have to prepare on a staff or in another administrative support role. Some of the guidance here is supported by guidance on naval correspondence covered in chapter 2.

STAFF WRITING GUIDANCE

"It's much harder to write the one-page for the four-star than the ten-page original document."

—NAVY CAPTAIN

Officers assigned to a staff will find themselves writing for many situations. Three aspects of staff writing deserve special consideration: the purpose, the person writing, and the boss or position of the commander or flag officer (for example, a squadron admiral or the Secretary of the Navy [SECNAV]).

THE PURPOSE OF STAFF WORK

The Navy has a longstanding saying: "Staffs exist to serve command." What does that have to do with writing? Purely and simply, on most staffs, especially shore staffs, the major result must be good "staff actions": effective directives, correspondence, plans, and other written documents signed out at the top level of the command.

For example, at OPNAV, the ultimate aim of the staff work is legislation and support for the Navy. Consequently, what counts the most there is what comes out the top, that is, what documents are signed by the Chief of Naval Operations (CNO), the vice chief, or one of the deputy chiefs. Almost everything an action officer does (phone calls, staff legwork, briefings prepared or attended, countless emails being read or sent, briefing packages put together, paperwork revised and proofed, and other duties) should contribute to this primary end, to what decision makers sign off or otherwise effect. Otherwise, all your painstaking work has no real effect.

The situation on a fleet staff differs somewhat. The goals are much more oriented toward operational requirements—what the ships in that fleet need to do their mission—than policy and legislation. Fleet-staff members must often focus much of their time on liaising with the ships; on helping keep the ships, aircraft, or other equipment in their command operational; and on ensuring they fulfill their operational missions.

As vital as it is, this liaison effort is almost always secondary to the primary need to support the boss's decision making. This priority becomes more and more evident as you climb the ladder of staffs in the chain of command. The higher you go and the larger the staff, the more the balance shifts away from operations and toward policy, and thus the more a staffer must concentrate on the research, the coordination, and especially the paperwork that can get that policy effected.

Clearly, the difference in staff size and staff nature will determine the degree to which a staffer must direct attention to such policymaking documentation. But in most cases, and especially on large shore staffs, wise staffers will focus their attention on the documents that come out at the end of the paperwork chain. Whether in electronic form or on paper, these packages ideally will be pure or complete staff products, having been so well researched and reviewed—in other words, so well filtered by good minds with wide understanding—that they are clearly and evidently the very best answers to the problems at hand. If staffers have done this work well, all that commanders have to do is to sign the documents for their decisions to become realities.

Yes, prior research, paperwork, formal briefings, endless emails and phone calls, and often off-the-record negotiations will have prepared the way for decisive documents of this sort. Then, afterward, the staffer must often put together directives, letters, briefings, and many emails to implement decisions. Both in formulating arguments leading to a decision and in providing a means to carry it out, the staff officer has much more to do than just writing the position paper, letter, memorandum-for, or other critical documents at the heart of a staff package. Still, such documents are the end results and goals of all primary staff work.

WRITING AS A STAFF MEMBER NEW TO A STAFF

Many service members reporting to a staff for the first time express frustration with their new duties. They often arrive after one or more operational tours. Indeed, many naval officers first report to staff duty as commanders or captains. Whatever their ranks, most have had little preparation for the kind of writing they now have to do. And many have a difficult time getting up to speed. For these new staffers, the greatest difficulty is usually the writing itself. Indeed, if you're going to get anything accomplished, you'll have to do it at least in part by the written word. To many, this requirement will seem a major obstacle.

One reason for this "staffer's shock" is the greater formality that may be required, especially at higher levels. For example, although the Navy and Marine Corps have many standard formats for staff documents, these are sometimes ignored in the field. Informal procedures or in-house templates often take over—a few notes appended to a letter sent forward for signature, a brief phone call that takes the place of writing. At major staffs, however, you can't succeed with informality.

For example, when a Marine lieutenant colonel first reported to HQMC, he found that, hard as he tried, he couldn't get anyone to pay any attention to the content of what he wrote until he got the format right—whether that of a point paper, a position paper, a briefing memo, or others. Because he had never run into the need for briefing on paper before, he found this situation highly frustrating. Others have had similar experiences. Navy writers have long operated by finding an old document of the type they now have to write and then copying its style. Yet such an approach is hardly optimal and does nothing to give a person the larger picture of good staff work.

WRITING FOR THE LEADER'S CONTEXT

Staff writing involves many audiences, but the most important are those empowered to make decisions. Of course, there are many decision makers, even several layers of them. They are typically highly knowledgeable but very busy individuals, people capable of absorbing facts and making decisions quickly. A former vice CNO recalled having literally no time to himself. He spent eight to ten hours attending briefings or meetings—or preparing for them—five days a week. Each week these left him with about four hundred packages of paperwork requiring action. All of them had to be handled after regular working hours. When could he handle all those documents? This vice chief began at 0630 on Saturday.

If you are writing for a busy senior officer, prepare the document to make review and decision making (approval or disapproval) easy. Imagine your report, memo, or email being just a few of those four hundred pages and consider how to best write for the perspective of that leader. All of the structural and style advice in chapter 1 applies here.

FIVE PROCESSES FOR STAFF WRITING

In addition to the general principles discussed in chapter 1 for identifying your content, structuring a document, and making the writing style clear and audience friendly, good staff writing involves several detailed processes, some of which overlap with these. Mastery of each process is a mark of the effective staff officer, whatever type of staff you serve on—Navy, Marine Corps, or joint. Described briefly here, the five basic processes are researching, writing the basic document, condensing the writing, assembling action packages, and coordinating one's staff work with other agencies. Some of this advice amplifies and complements the general writing guidance in chapter 1.

1. Conduct Thorough Research

A letter, email, message, or oral command (usually called a "tasker") typically initiates staff research. Several hours, days, or even months of work may follow, either on your part alone or as a team effort. Clearly, spending this time well is important. Start by carefully focusing your research.

First, consider the problem. Analyze it and restate it for yourself if necessary to clarify the problem and define its scope. Sometimes your tasking will be incomplete and vague. Check widely with authorities and clear thinkers to make sure you have the larger picture clear before putting out a Herculean effort. And don't hesitate to redefine your original statement of the problem if you find your initial information doesn't fit the facts.

Ensure you know exactly what your boss wants. As one former OPNAV staffer commented: "How many times have staff officers busted their bums on a package only to have the boss say, 'That's not what I asked you to find out'? Ask the boss as many questions as he or she can tolerate when you're given the tasking."

Once these expectations are clear, limit your scope so you don't spend weeks on a problem you can't affect anyway. And make sure the effort you expend is worthwhile. If your problem is ship stability in the North Atlantic, don't spend much time on the free surface effect in the ship's toilet bowls and coffee pots.

Develop (and state) criteria for solutions. Sometimes the tasking memo or other order that has initiated your staff work will give the criteria for a solution; sometimes they will be obvious. But if you take care to formulate the criteria in writing, you will be sure to focus on the particular cruxes of the problem—the meaty, difficult parts.

Gather data and seek additional information as necessary, following up leads your first research has suggested. Use your own wits as well as the collective experience of your coworkers to define the best sources. Interpret the data, with an eye to solutions, and work out the implications of the information you've researched. Organize the information into possible solutions to your problems.

Identify and evaluate the alternatives. Use the criteria you've developed in your evaluation. Remember these three classic tests of any proposed answer to a staff problem:

Suitability. *Will it in fact solve the problem*? Scores of planners have stumbled because, having designed a weapons platform, discovered a new tactic, or worked out a new system of personnel motivation, they find that it doesn't solve the problem they faced originally (and they should have known that shortcoming beforehand).

Feasibility. *Can it actually be done*? A brilliant concept is one thing; working it out in practice is another. Do all the spadework to see if a great idea is practicable.

Acceptability. *In the overall picture, is it worth the cost*? Many solutions may meet the first two tests, but the question may become one of cost. Cost can be measured in terms of money, equipment (including ships, tanks, or aircraft), lives, or troop morale, and energy; it can be measured in moral, social, or political terms as well.

Incidentally, the ancient Greeks developed a series of questions that is very similar to these three tests. They suggested the following three questions could be asked of any proposed course of action: Is it possible? Is it expedient (that is, what good will it do)? Is it just? (Don't forget to ask that last question.)

Many a solution will pass one or two of the tests outlined above, but only a solution that definitely passes all of them is likely to be rock solid. Realize also that sometimes the answer you come up with will meet all the criteria but won't mesh with the way your boss thinks. At other times, your boss will like it, but his or her superiors won't—it isn't "what they want to hear," as the saying goes. You can't always give superiors answers they will be comfortable with. You should, however, get a feel for the political climate before making recommendations and try to measure the costs of fighting for any particular solution or recommendation.

2. Prepare a Draft for Peer Review

Having thoroughly researched the problem and determined the best solutions, you should compose a complete draft of a report. Whatever the format and whatever its stipulated length, the best way to start is by writing up the report thoroughly, recording the problem, assumptions, criteria, evaluation of alternatives, and recommendations, as mentioned above. Then you can assess your thought processes and look for holes in both data and logic (a quick sanity check) before you put it all up for review.

When you have your thoughts in a more or less presentable form, have some of your colleagues review your work. This step is especially important for new staffers. But even old hands who are experienced authors of Pentagon staff work, *Proceedings* articles, and speeches for senior officials depend on reviews by knowledgeable colleagues for feedback somewhere in the process. Often such review will prompt you to conduct more reflection and even more research. If that extra work results in a better product, it's usually worth it.

Periodic reviews at early stages of an especially long or complex project can reduce wasted effort later. In any case, on the basis of all reviews, rethink your concept and reformulate it as needed.

3. Revise to a Smooth Final Document

Once you've done all your research, decided on the best course of action to recommend, and written up the whole process in logical order and thorough detail, you must then write the brief memo, letter, or point paper that will get that action put into effect.

In others words, having the perfect solution and even laying it out in perfect clarity and detail isn't enough—you must also convince your busy boss to adopt it. So, instead of that ten-page research report that you originally wrote, you need to condense it into a one-page briefing memo. Why so short? Again, consider the predicament of the boss for whom you write. As one general commented: "To work the problems of the central battle within the restrictions of the realities, I need less information. But every piece of the less has to be pure. You need to synthesize, condense, strip out, boil down like a good newspaper editor." How do you boil it down? Condensing your ideas can be difficult, but good editing skills can help greatly. Summarized later in this chapter are some of the most helpful editing techniques for staff work.

4. Prepare a Correspondence Folder

Having written the briefing memo, letter, point paper, or other documents that the particular problem requires, you must assemble the correspondence package.

Gather together all the paperwork pertinent to any staff action: the document to be signed or approved, the briefing memo, and all necessary explanatory material.

Pay special care to putting the folder (which might well be digital) together well. No matter how well you write, your boss will probably send your package back to you if you haven't included the basic references (or excerpts from them) or if you haven't made the required number of copies, used the right forms, or committed other such errors. This package must be technically correct in every detail.

5. Coordinate: Learn How to Get Paperwork through to the Top

Finally, there is an art to routing and negotiating a package through a major staff. No matter how good the research, how pointed and cogent the writing, and how technically correct your correspondence package, you will accomplish nothing if you don't send that package to the right people, in the right order, for the right kind of comments or reviews.

Realize that at big commands (like OPNAV and HQMC) the boss will look for the coordination signatures, or "chops," before signing. Without the right chops, the substance will be irrelevant; the package will be returned to you unsigned. On the other hand, if all the chops are done, the package may be signed and sent on immediately.

The following are a few pointers on good coordination, drawn from interviews with various Navy and Marine Corps authorities and from the *Correspondence Manual*.

Consider Who Should Sign the Briefing Memo

The briefing memo (or "routing sheet") accompanies a document through coordination. If the matter is very important, the boss of your office, or "shop," had better sign this document.

Decide Who Needs to Coordinate

Some staff members must see your package before you have it signed, while others can be informed afterward. If unsure whose chops to obtain, ask an experienced coworker or consult an organization chart. Many commands have desktop guides that spell out the required coordination. A large staff will often have a "Secretariat" (that's the name used at OPNAV) designated to review your chops before your package goes to the boss. Seek out this office and use it regularly to learn all the nuances of "chop chains."

Establish Your Own Network

Make contacts with people in other offices in your building with whom you can talk about your packages. Find out to whom you can send material and learn who might shepherd your package through their shop in a hurry, if that's ever necessary.

Vary Coordination by Reference to the Particular Matter at Issue

Consider chopping by phone if the matter is brief and routine. If revisions are likely, you might want to coordinate in the drafting stage.

Plan Out a Strategy for Getting Signatures

The sequential chop is the customary method. In this process, you simply send an email package of files (or a traditional paper correspondence folder) sequentially from office to office, indicating on the routing sheet who gets the folder and in what order. Eventually, the last one on the chain will return it to your office.

The shotgun chop can be useful if time is short. With this method, you "shotgun" the document or package to many offices simultaneously and then summarize the responses on a briefing sheet to preface the package as it goes up for signature.

But the shotgun chop can be a two-edged sword, creating extra work. In a sequential chop chain, each succeeding office gets to review the preceding chops before acting on the package. Often, this review mitigates or tempers suggested revisions. With a shotgun chop, writers respond without any knowledge of the inputs of other offices. The drafter must then reconcile all responses before going forward and may have to get a "re-chop" from one or more people. It isn't good staff work to send a batch of conflicting responses forward, leaving it for the boss to iron out the differences.

There is at least one other method of getting chops. If you are very short on time, you can sometimes hand carry a package through offices personally, indicating to all individuals the material specifically pertinent to them. If you are routing documents for signature electronically, you can communicate personally with each signer to ask them to review and approve quickly. This way you can keep superfluous changes to a minimum and often get a document through the process very expeditiously.

Track Your Package and Keep It Moving

Your job isn't done when you send off your package. Correspondence packages can collect dust for weeks in some offices. Although the best ones have a tickler system that keeps track of when correspondence comes through and when (and if) it moves on—in OPNAV the "tasker" system keeps things flowing—not all of them do.

Know where your package is, and keep it moving along.

An LDO on a major staff once pointed out: "Sometimes a package will wait too long for somebody's signature and will get to CNO and SECNAV months after its origination. Then they'll send it back for a rechop, and it will have to go back through all the very same offices. This will double the time before it's signed off."

Common Documents for Staff Writing

You can expect to write several kinds of documents when serving on Navy and Marine Corps staffs. The various types of correspondence also common on staffs and elsewhere—emails, letters, and memos—are discussed in chapter 2. Guidance on naval messages appears in chapter 4. Other documents not covered here that you may encounter will likely follow the same principles of structure and clarity discussed below.

BRIEFING MEMO OR ROUTING SHEET

"At the senior-officer level, a good route sheet or briefing memo is invaluable."

—MARINE CORPS COLONEL

In OPNAV a briefing memo is sometimes called an "executive summary memorandum." In HQMC paperwork usually moves by a "routing sheet." Joint commands use many names, including "decision paper" and "summary sheet." Whatever its name, the ability to write one effectively is an essential ability for any staffer, whether officer, senior enlisted, or civilian. A briefing memo accompanies another document or some issue to explain its content and context for the senior officer. Figure 3.1 provides some guidance on how to create a strong briefing memo. We have put the advice in the form of a sample briefing memo with common subsections, though specifics of formatting may vary at different commands.

Figure 3.2 depicts a particular type of memorandum-for (see chapter 2) that functions as a briefing or decision memo and is sometimes called the "executive summary memorandum." This example is fictional but is based on actual OPNAV documents. Note that some writers prefer numbered paragraphs in the discussion section rather than the "bullet style" used here and in figure 3.1.

POINT PAPER

Point papers are good ways to press forward recommendations in a direct and objective way. Instead of writing a letter or other document for signature and then attempting to persuade seniors to buy your approach, you can write a point paper

---

FIGURE 3.1. *Briefing Memo Guidance in Briefing Memo Format*

Date: 30 Nov 23

Subj.: WRITING THE BRIEFING MEMO OR ROUTING SHEET

ISSUE: A briefing memo is an explanation sheet—similar to a cover memo or email in the civilian world. It's a piece of paper that explains any letter, memorandum, instruction, or other document that needs to be signed by the boss and issued. You can also use it to brief the boss even when there is no outgoing correspondence.

RECOMMENDATION: Make clear everything you are asking regarding the main document or issue.

- State the central issue, your primary recommendation or point, clearly near the beginning.
- Keep your memo to one page.
- Write in brief, logically ordered bullets or numbered paragraphs.
- Avoid duplicating what is in the main document.
- Do not simply write, "Sign the attached correspondence."

BACKGROUND: The briefing memo also gives context for the main document: a concise history of the package with supporting and explanatory remarks. In addition to summarizing the primary recommendation, use the briefing memo to

- explain what you're proposing and why,
- discuss any rejected alternatives, and
- address why the package is late, if it is.

COORDINATION: Ensure you obtain all necessary chops (reviews and signatures). Mention the important coordination you have obtained if it's not clear in the memo or correspondence. For instance, "The proposed response was coordinated with 04B, and it was chopped by PERS-9."

Signature:
Action Officer, LCDR, USN
Office Code & Phone: 03A/x5-5555

---

seeking a decision and wait to implement what your boss decides. As one Navy captain remarked: "With the point paper you create a grenade with the pin out, but without requiring anyone in the chain to sign it. It's an excellent way to direct things."

Point papers serve many contexts and purposes. For instance, you can use one to bring up issues in conferences, to help develop policy, to help resolve differences between offices, or to prepare senior officers for appearances before important bodies, such as congressional committees.

In most of these documents, you are trying to talk your audience into something; therefore, unlike a briefing memo or other staff action papers, you can often leave the recommendation until the last. As long as you keep the point paper to one page and use clear headings, the audience can skim the document and find your "bottom line" very quickly. Figure 3.3 is a model point paper, which follows the format found in the *Correspondence Manual.*

FIGURE 3.2. *Memorandum Used as a Briefing and Decision Sheet*

DATE: 16 Apr 24

MEMORANDUM FOR Chief of Naval Operations

From: RADM Byrhtnoth
        Director, Intersubair Operations
        Prepared by: CAPT Breton, Head
        Subair Warfare, NXXX, 697-xxxx

Subj: PERSONAL FOR MESSAGE RESPONSE TO COMSECONDFLT–ACTION MEMORANDUM

PURPOSE: To obtain CNO's release of PERSONAL FOR message in response to

COMSECONDFLT message discussing aircraft "Dive-Under" attack maneuvers and fleet training.

DISCUSSION: COMSECONDFLT (RADM C) sent PERSONAL FOR message (attachment 1) to inform CNO of benefits of fleet training during mock attack by PELICAN model Dive-Under aircraft. These aircraft are usually armed with subcruise missiles.

- Message included detailed comments by CO USS FAMOUS CITY upon the effectiveness of MK-99 lightweight SUPER-DUPERs in encountering intermittent subair threats but urged extensive real-time training in the use of such countermeasures.

- There were no controversial issues. CNO asked for simple response.

PERSONAL FOR response prepared; at attachment

RECOMMENDATION: CNO sign message release form to right.

ATTACHMENTS:
1. COMSECONDFLT message 141415Z APR 24
2. CNO PERSONAL FOR message response

FIGURE 3.3. *Point Paper Guidance in Point Paper Format*

POINT PAPER

Rank and Name
Staff Code, Phone Number   18 Dec 23

Subj: USE OF POINT PAPERS

BACKGROUND (or PROBLEM)
Point papers are a good means of stating background, ideas, and recommendations in a relatively formal way for the consideration of the command. Use a point paper primarily to direct the attention of seniors to an issue or problem and to seek a solution.

DISCUSSION
- Keep to one page in most cases; use tabs for additional material.
- Be factual and objective.
- Keep the language simple. Explain all technical terms or unfamiliar acronyms the first time you use them.
- Don't make the point paper so detailed that significant points are lost in minutiae.
- Indicate who concurs or does not concur.
- For classified papers, follow markings found in *Correspondence Manual*.

RECOMMENDATION(S)
State recommended actions. Be brief but specific, outlining who, when, where, how much, etc. List options, if desirable, but always make your choice clear among them.

Figure 3.4 is an example of a strong point paper from the 1990s. A civilian manager at Military Sealift Command, Pacific had identified a problem. His naval commander then asked him to outline the problem in a point paper that could be used in a conference with CINCPACFLT. Through the discussions on the topic as described in this point paper, MSCPAC obtained its objective.

Figure 3.5 is another example of a point paper prepared on board ship some years ago for use at a conference at an operational staff. It follows the format outlined above except that it uses numbered paragraphs instead of bullets.

FIGURE 3.4. *Example of a Strong Point Paper*

Logistics Directorate
MSCPAC N4
(123) 456-7890
April 1995

Subj.:   Transferring Material Handling Equipment (MHE) to MSC
Ref.:   (a) COMNAVSURFPACINST 4100. IE
             (b) COMNAVSURFPAC San Diego 2113172 DEC 94

Encl.:   (1) COMNAVSURFPAC San Diego 0505172 APR 95

BACKGROUND: USS FLINT is preparing for a "hot transfer" to Military Sealift Command. Previous ship transfers to MSCPAC, including SAN JOSE, MARS, and NIAGARA FALLS, have included their onboard MHE assets. However, FLINT's MHE had been scheduled by COMNAVSURFPAC to be offloaded for ultimate distribution to other ships. Following a formal request for reconsideration by MSCPAC to COMNAVSURFPAC, this process is currently pending further disposition instructions from NAVSUP and SPCC. Encl. (1) applies.

DISCUSSION: Current guidance provided by COMNAVSURFPAC in reference (a) applies only to the deactivation and subsequent retirement or sale of USS ships. Reference (a) briefly mentions MSC ship transfers and states that "removal of equipment and material is generally prohibited except as authorized by the Type Commander *and coordinated with COMSCPAC*" (emphasis added). However, reference (b) directed the offload of MHE from FLINT to FICP San Diego.

Actual ownership of MHE is the primary issue. Reference (b) defers to NAVSUP and SPCC for FLINT's MHE disposition instructions. Seemingly, a more logical choice would be for these instructions to originate from the owners of afloat MHE assets—Fleet Commanders. MHE shortfalls may adversely affect operational readiness and/or financial planning for the CINCs. For example, replacement of FLINT's MHE represents an unplanned shortfall of 1.3 million to MSC and up to 2 years for new MHE units to process through existing contracts at SPCC.

RECOMMENDATION: CINCPACFLT establish a policy for COMNAVSURFPAC, NAVSUP, and SPCC which clearly states that MHE remains onboard for any "hot transfers" or deactivations where eventual transfer to MSC is anticipated.

FIGURE 3.5. *Point Paper as Conference Preparation*

Point Paper for COMCARDIV STAFF   DATE: 2 DEC xx

ORIG: USS CARRIER

DRAFTER: LCDR R. ENGINEER

SEA/SHORE DUTY ROTATION FOR NONNUCLEAR-TRAINED MACHINIST'S MATES

PROBLEM

Sea-duty obligation for E-7 through E-9 machinist's mates has been increased to 60 months.

DISCUSSION

For the second time in less than three years, the sea-tour length for nonnuclear machinist's mates has been increased. The total extension has been 24 months. The results are as follows:

    a. Shore-tour planning has been superseded because detailers and career counselors do not have an accurate list of options to discuss.

    b. Family planning in regard to PCS moves, retirement options, and future education plans for children is in jeopardy due to uncertainty about career options.

    c. Personal career planning—to stay in or get out—is being affected: many senior qualified personnel are choosing to leave the service.

Personnel on board USS CARRIER are confronting a sea tour with not just one or two six-month-plus deployments, but possibly three. For senior enlisted, the options are limited—either retire or deploy. Prior to this extension policy, some had decided to stay in and contribute at a shore facility that could use their technical expertise and knowledge.

RECOMMENDATIONS

    1. On a case-by-case basis, review the sea/shore rotation dates of all senior machinist's mates. Send men whose shore-tour length (two years) has been fulfilled back to sea to relieve those men whose tours have been extended.

    2. Reestablish SRB for machinist's mates at a level that will support the required retention level. Many predicted the current problem when SRB was reduced several years ago.

TALKING PAPER OR TALKING POINTS

Another staff document similar to a point paper is a talking paper, also called "talking points," so named because you usually prepare it for someone to use while speaking in informal circumstances. These circumstances might be interviews with visiting officials, informal talks to groups, or visits with the media, among others. Keep the talking paper brief and simple. Figure 3.6 provides guidance in the form of a sample. Your command may have a template or a set layout you should follow as well.

Figure 3.7 is an example of a talking paper, adapted from an actual one once used at BUPERS.

FIGURE 3.6. *Talking Paper Guidance*

Subj : HOW TO WRITE A TALKING PAPER

Originator's Name, Code/Phone Number   Date Prepared

BACKGROUND (or ISSUE)

- The name of the official for whom this paper has been prepared, the name of the meeting, etc.
- The event or situation that has brought this issue up now.
- Any other brief background needed.

DISCUSSION (or TALKING POINTS)

Use this outline as a memory aid in a meeting, or as an informal agenda. Also use it as a tickler to prepare seniors for meetings with important officials, such as senior Navy or Marine officials or Congress members. Tailor the talking paper to the boss's preferences, but in general follow these best practices:

- Include the most important facts.
- **Be concise** in wording.
- Use elements of information design (visual layout) to make key information easy to skim and find:
  - Subordinate points clearly
  - Use **bullet style**
  - Use **line and paragraph spacing** for easier reading
  - Keep a talking paper or talking points to **one page**
  - Consider putting **key terms, ideas, or data in bold** for ease of quick reference
- Say what to avoid talking about as well as what to talk about. Also note, if needed, who has been involved, who concurs, who does not concur.
- Mark classification (and paragraph classification) as required. See SECNAVINST 5216.5D.

RECOMMENDATION(S)

Include a "Recommendation" section if needed. If there are no recommendations, omit this section. Consider prefacing the talk by saying, "This presentation is for information only."

TRIP REPORT

If you travel for your work to attend a meeting or a conference or to accomplish a specific mission, you may have to write a report documenting your trip. Some people dislike writing these and see them as busy work, but a trip report can be an important and useful management document. Ideally, trip reports can contribute to long-range command strategy.

The CO of a Navy training command, for example, had the following strategy: "I send my people to get things to happen at meetings, not just to listen. If we're not playing, we're not going. The trip report isn't a drill—it is the purpose of the trip, the reason you went. It becomes a management tool that (through its recommendations) has an impact or gets things done, either in the Navy at large or at home in your own command."

FIGURE 3.7. *Sample Talking Paper*

TALKING PAPER

CDR S. F. Housing OP-999H/#5-4321

13 Apr 2022

Subj: SHORTFALL IN FAMILY HOUSING (MFH) AT BIKINI ATOLL

BACKGROUND

- This paper was prepared at OP-99's request, in response to an inquiry by Assistant Secretary of the Navy (Logistics). This issue may come up in a meeting between OP-99 and the Secretary next Tuesday, April 17.

TALKING POINTS

- MFH assets at Bikini are owned and managed by the Air Force.

- Navy owns no housing there.

- Average waiting time for personnel to get into housing is 8–10 months.

- Figures relevant to MFH at Bikini:

  – Total requirement (Air Force and Navy families): 900

  – Total of 392 Air Force & Navy personnel are on the housing waiting list

  – Current assets: 182 units

- Programmed for construction:

  – FY 23: 150 units (approved)

  – FY 24: 150 units

  – FY 25: 100 units

COORDINATION SUMMARY

- USAF point of contact is LtCol D. A. Quarters at 1-2345, who provided some of the above information.

A Navy captain remarked: "It's not worthwhile reading trip reports that list the lectures you heard. Trip reports seem to be used to account for your time— but no one cares about that. Give a sense of what you are discovering." And as a Marine lieutenant colonel commented, "Don't just say 'I came, I saw this, I talked to so-and-so'; make recommendations."

As the captain suggests, many staff members who attend meetings are unprepared to contribute significantly to the proceedings. Sometimes a pretrip report will be part of the process as well, documenting in advance why you are going or what you hope to accomplish. A report that requires staffers to outline their reasons for the trip, the people who will be there, the business they will conduct, and any controversial topics they expect to arise can help staff members learn to regard conferences as means to an end or as opportunities to make things happen. After such

---

FIGURE 3.8. *Trip Report Guidance*

15 Mar 2023

From: Senior Officer or Officers Who Made the Trip
To: Commanding Officer-via the Chain of Command
Subj: TRIP REPORT FOR [MEETING, CONFERENCE, ASSIST VISIT, ETC.]

Encl: (1) Agenda or itinerary
(2) List of attendees
(3) Minutes or other enclosures as pertinent

Trip Purpose: Record the objective of the trip you took. Why was the meeting held, and why did staff members go? What did they expect to get out of it? Whom did they expect to influence and why?

Highlights: Comment in bullet format on the outcomes.

- Whether the organizers achieved their aims.
- What major decisions participants made and what new information they issued.
- What milestone status on projects, proposals, or participants reported.
- Whether you achieved your objective in taking the trip.

Unresolved Issues:

- Newly discovered problem areas.
- Unresolved issues, including their current status.

Action Items or Recommendations:

- Items assigned to your command for action.
- Anything you committed your office or boss to.
- Your recommendations regarding ways of doing business, new developments or projects, and other potential application.

Opinions and Impressions: In this closing paragraph, comment on such matters as the overall success of the trip, developments in other areas, ideas about what the future holds, or other applicable thoughts for your audience.

Very Respectfully,
S. Officer

---

formal preparation, a posttrip report (like the guidance example in figure 3.8) can be a way to provide recommendations for action and changes in policy as well as to relay new information.

Whatever your command's strategy, remember to keep a sense of priority as you pen your final report. As with many other staff documents, make a habit of putting the vital points up front—at least on the first page—or the report simply won't be read. Leave for appendixes such peripheral material as topics covered at the conference and lists of speakers. The format of the trip report may simply be a standard memorandum, a layout specific to your command, or you may be free to design it yourself. Regardless, the content is most important.

LESSONS LEARNED

Lessons learned are naval problem summaries. Widely used on both operational commands and staffs, a lessons learned document reports on difficulties in recent operations, exercises, inspections, and many other evolutions. These papers usually describe problems that have already been solved, recording a solution alongside each specific problem. (For those yet unsolved, staffers typically compose point papers and then follow them up with Plans of Action & Milestones [POA&Ms], briefing packages, proposed revisions to directives, and other items.)

We record lessons learned both to keep present commanders informed and to guide personnel in the future. Not only can individuals forget from one evolution to the next, but given the rapidity of personnel transfers (or wartime casualties), the next month's or next year's evolution will often see different personnel in key positions. These new people will badly need guidance to rely on and will not want to start from scratch.

One wide use of lessons learned during peacetime is to record problems solved during fleet or Fleet Marine Force exercises, as in "Red Flag Lessons Learned," "Exercise Provide Promise Lessons Learned," and others. But we write these papers on many occasions other than exercises—after major inspections or after standard training evolutions, for example. In such cases their purpose is to guide those who must prepare for the next major training cycle or other regularly occurring exercises.

Furthermore, a ship or unit "chopping" into a new operational area (a destroyer reporting to the Indian Ocean, for instance) will often receive from the ship or unit it is relieving an after-action report itemizing lessons learned. In this instance the document would fit into the standard naval relieving ceremony as a kind of turnover file. Comprehensive lessons learned are often written on major naval events to discuss strategy, tactics, successful employment of new weapon systems, and the like. For a good ten-page example of one of these studies, see "Lessons Learned on the Falklands War" in former Secretary of the Navy John F. Lehman Jr.'s *Command of the Seas* (1988). Lehman himself drew on two book-length lessons learned for his discussion.

Write lessons learned often—whenever specific experience has taught something you (or someone at your ship or station) may need to know in the future. Make sure to incorporate these documents into your turnover file. Whatever the topic, remember not to talk just about what went wrong; also discuss what went right. Sound procedures, reasonable rules not to bend, planning ahead that proved right on target—all this information is as important as "problem-solved" commentary.

FIGURE 3.9. *Lessons Learned*

Lesson Learned

TOPIC: Problems Involving Communications Security

BACKGROUND: Drawing on the exercise plan and additional guidance received during pre-briefing, units employed communications-security procedures (mainly shackling) from Day 1 of the exercise.

DISCUSSION: Such employment of communications security is doing more harm than good. During the exercise, grids were coded or decoded incorrectly, important messages were delayed because of coding and decoding requirements, CEOs were not where they were needed, aircraft were on the wrong frequencies, etc.

RECOMMENDATION: For a short Reserve exercise, we should waive shackling. It is invariably done improperly, wastes time, confuses personnel, and therefore delays other training objectives. In addition, on D-day the enemy is likely to know exactly where you are, and shackling your grid coordinates from a known position allows him to break your code.

The example in figure 3.9 is excerpted from a lessons learned drawn up following a Marine Corps Reserve exercise. It follows a widely used format, first describing the background, then presenting a brief discussion, and finally making a recommendation. As always, format can vary with command and context.

Another good example of lessons learned is a Naval Surface Force, U.S. Atlantic Fleet (SURFLANT) document (in memo-for-the-record format, discussed in chapter 2) adapted from the original and presented as figure 3.10. The author was officer in charge of a change of command. He recorded his lessons learned and placed them in the local "Change of Command" folder so that future ceremonies would also go well.

PLAN OF ACTION AND MILESTONES

The POA&M is a planning document used widely throughout the Navy and Marine Corps as well as in large organizations elsewhere. It is exactly what its name suggests. It is a plan of the steps required to complete a project, inspection, or other evolution, with milestones for each step. Virtually any organizational activity can formulate a POA&M, but it's especially appropriate in cases where responsibility for action involves many different people, divisions, or departments.

POA&Ms are perhaps most widely used in the surface Navy, both on staffs and on ships. Staffs use them for instituting new programs or getting initiatives off the ground. Ships use them to guide preparations for myriad inspections and to schedule the correction of discrepancies after inspections. You can also use POA&Ms to schedule change-of-command ceremonies or to plan deployments and training exercises.

FIGURE 3.10. *Lessons Learned in the Form of a Memo for the Record*

5 Jan 20xx

                    MEMORANDUM FOR THE RECORD

Via: (1) N3
        (2) 002
        (3) 02

Subj: CHANGE OF COMMAND LESSONS LEARNED

1. This memorandum addresses the planning for and execution of the 20xx COMNAVSURFLANT Change of Command. In general, the event was a success; this memo addresses areas needing special effort in paragraph 4.

2. As background, a successor to VADM was not known until very late in the game. The same was true of the actual date for the change. As a consequence, specific planning was delayed until about four weeks prior to the event. This time compression had the greatest effect in the expected areas: Invitations printing/mailing, program printing, etc. VADM was frocked after the programs were printed, which necessitated printing them again—SHIP NAME did it in three days.

3. Specific responsibilities were assigned by COMNAVSURFLANTNOTE. Three TEMDU officers were assigned to the project and were able to devote almost full time to it, which blurred some lines of responsibility. This minor disadvantage was outweighed by the clear advantage of full-time help.

4. The "hard" areas included the following:

    • **Communications**: Although ultimately resolved, a lack of hand-held comms was an issue. Need to resolve early, identity requirements, and obtain the radios as soon as possible.
        – Portable comms didn't work well from LP-1.
        – Don't depend solely on NAVSTA to provide hand-held comms. Task one of the Groups early to provide units from ships' assets.

    • **Ceremony timing**: The ceremony was scheduled to start at 1000, which meant that the CNO arrived at 0959. This meant that the principals arrived sometime before that, causing the forward brow to be secured in preparation.
        – Flag officers attending a 1000 event will arrive at 0955.
        – Some did and had to use the after brow.
        – Philosophical Question. Does the ceremony start with the benediction, or when the principals arrive? The issue should be broached next time.

    • **VIP transportation**: N7 handled this perfectly, but it took a major effort.
        – It's a moving target, but identify VIPs who need transportation, then add five and order that number of cars.
        – Ask for a list of escort officers from the staff (ACOSs) early. One per car per run, basically.
        – Drivers were staff CPOs—trained by N7—worked well. Lots of practice runs help.
        – Attention to minute detail required in this area. Fertile ground for major problems.
        – VIPs (aside from principals/principals' families) were out-of-town flag officers who flew in for the day.

    • **Standard change-of-command items**: Don't overlook mundane items such as UNITREP, releasing signatures, security badges, etc.

    • **Reception**: Nail down as early as possible. Provide options. Sensitive issue—personal money involved. No government funds should be expended on solely reception items. N7 will explain.

    • **Uniform issues**: Make the uniform requirements clear for every event and then get the word out.

5. Expect full cooperation in the overall effort. Navy Regs and the Protocol Manual provide lots of guidance as questions arise. Start as early as possible.

                                        Very respectfully,
                                        Commander, USN

The POA&M is very simple in concept and appearance. You can design yours in different ways, but usually a command will have a template. Typically published in an instruction or other official directive, the basic plan consists of a schedule of action to be taken and a designation of the cognizant official for each action. In print the document usually consists of three columns, labeled Action Item, Action Individual, and Due Date; these simply designate what needs to be done, who has to do it, and by when.

This simplicity may be misleading. Besides taking care to be accurate, you must be very thoughtful and farsighted to make the POA&M a workable document, one that makes allowance for other unit evolutions that may affect the plan. Even after you design it, it's not set in concrete. As one officer on a DESRON staff cautioned: "The POA&M never happens as you schedule it. Crises are always coming up to interfere with it." You need to revise it often. Indeed, good as it is in concept, a POA&M will be absolutely useless in practice unless you check progress frequently, identify and overcome obstacles, and hold people to the stipulated deadlines. Without such rigorous management, the plan will simply be ignored.

POA&Ms have special value for commanders, who can gauge the command's progress toward a goal simply by glancing at the chart of goals and accomplishments it comprises. Because a superior will often judge a department's or a division's progress by looking at a POA&M, you can put yourself on report by being less than farsighted in working one up. Make sure you have a reasonable chance to complete any tasking before you assign yourself responsibility to do it. Have the same consideration for your subordinates when giving them responsibilities. In short, formulate a POA&M with discretion and evaluate it with understanding.

Tips for composing a useful POA&M:

- Be very detailed with assignments. Separate complex activities into a number of individual steps.
- Assign responsibility for each step by billet or code to one specific individual.
- Assign reasonable due dates.
- Update the plan periodically, adjusting the dates and responsibilities according to changes in schedule, available personnel, etc.
- Keep the time frame of the POA&M to less than one year. The longer the time and the greater the detail, the more unwieldy the plan becomes. It is likely to become meaningless if stretched too far into the future.

Figure 3.11 is a POA&M used aboard a surface ship for scheduling the qualification of Enlisted Surface Warfare Specialists (ESWSs). POA&Ms on staffs are often

| FIGURE 3.11. *Sample POA&M for ESWS Qualification* |

Milestones. PREBLE has established two weeks before outchop as the deadline for 100 percent ESWS qualification of eligible petty officers. Adhere to the following schedule in accomplishing that objective:

Action Item	COG	Due NLT
a. Promulgate list of Chiefs/POs aboard	XO	01 May
b. Revise Monthly PQS reports to track progress toward ESWS quals	XO	01 May
c. Conduct review of ship's program, and establish reporting requirements	CO	15 May
d. Promulgate updated ESWS Qualifiers list	XO	15 May
e. Develop timeline that reflects time remaining vs. number of qualification points completed, in order to qualify no later than 15 October	XO	15 May
f. Develop bank of 500 questions for use in ESWS qualification exam. Publish question list for crew's review	SMCS	01 Jun
g. Commence review of ship's program and ship's qualifications	SMCS	15 Jul
h. Identify individuals who have not made acceptable progress, and report delinquents to XO	CO/XO	01 Aug
i. Establish after-hours schedule of instruction for delinquent personnel, and promulgate schedule to crew	SMCS	01 Aug
j. Review ship's program and status of crew's qualifications	CO/XO	15 Aug 31 Aug 15 Sep 30 Sep 15 Oct

more complex (sometimes they have many sections), but otherwise they are usually similar in basic layout.

DIRECTIVES

"I had to learn that, particularly as a junior officer, it wasn't about writing as an individual but writing for the command. You had to learn to speak in the generic: 'This is the USS Maine *speaking.'"*

—NAVY LIEUTENANT

The following passage, the purpose paragraph from a sample instruction, is taken from a guide on how to write instructions and notices:

Purpose. To produce forth a guide by which originators may formulate instructions and notices following the provisions of references (a) and (b). This instruction covers the procedures that originators will carry out in writing directives in the Navy Directives System.

What's wrong with this paragraph?
- Both sentences say the same thing.
- The sentences contain unnecessary words:
  - "the provisions of" could be omitted in the first sentence without loss.
  - "in the Navy Directives System" could be omitted in the second.
- Cumbersome words and phrases obscure simple ideas:
  - "To produce forth a guide" is an awkward way to say "to guide."
  - "formulate" is officialese for "write."
- In sum, all four lines could be revised to *eight simple words* with no loss of meaning:
  - *Purpose*: To guide originators in writing directives.

This paragraph should have been an outstanding example of how to start off a directive. Instead, it is an example of how *not* to begin. Unfortunately, many directives read more like this poor example than like our revision.

"Review your directives often. I made a policy of reviewing all my instructions within nine months of coming aboard, and in one case I was able to cut them down to a fifth their original size. Once they had been made manageable in this way, I could insist my people knew what was in them."

—CAPTAIN, NAVY

The CO, admiral, general, or SECNAV may sign various command instructions directing action and policy, but those individuals do not do the actual writing. You may well find yourself composing—or more likely updating—a directive that needs to be acceptable to the senior official who must sign and clear it for a large number of people to follow.

Written directives—instructions and notices in the Navy, orders and bulletins in the Marine Corps—are even more vital in the armed services than in other big organizations because Sailors and Marines transfer from one outfit to another so often. This regular turnover leaves little corporate memory in units or commands, making the guidance of various directives all the more valuable. Yet if directives are too dense, too long, or too complicated, they will not be read, and your people will "fly by the seat of their pants."

THE NAVY NOTICE AND THE MARINE CORPS BULLETIN

The Navy notice and the Marine Corps bulletin are distinct among directives in having short-term authority. Navy notices, for example, cannot remain in effect for longer than a year, with most lasting six months or less. Because of their relative impermanence, they are best used for one-time reports, temporary procedures, or

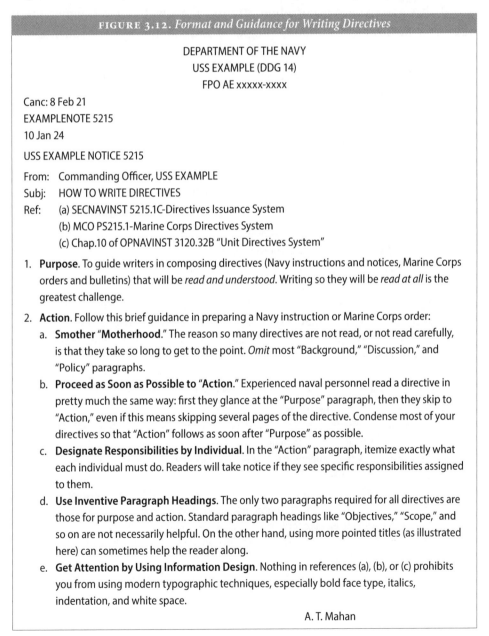

FIGURE 3.12. *Format and Guidance for Writing Directives*

DEPARTMENT OF THE NAVY
USS EXAMPLE (DDG 14)
FPO AE xxxxx-xxxx

Canc: 8 Feb 21
EXAMPLENOTE 5215
10 Jan 24

USS EXAMPLE NOTICE 5215

From: Commanding Officer, USS EXAMPLE
Subj: HOW TO WRITE DIRECTIVES
Ref: (a) SECNAVINST 5215.1C-Directives Issuance System
        (b) MCO PS215.1-Marine Corps Directives System
        (c) Chap.10 of OPNAVINST 3120.32B "Unit Directives System"

1. **Purpose.** To guide writers in composing directives (Navy instructions and notices, Marine Corps orders and bulletins) that will be *read and understood*. Writing so they will be *read at all* is the greatest challenge.

2. **Action.** Follow this brief guidance in preparing a Navy instruction or Marine Corps order:
   a. **Smother "Motherhood."** The reason so many directives are not read, or not read carefully, is that they take so long to get to the point. *Omit* most "Background," "Discussion," and "Policy" paragraphs.
   b. **Proceed as Soon as Possible to "Action."** Experienced naval personnel read a directive in pretty much the same way: first they glance at the "Purpose" paragraph, then they skip to "Action," even if this means skipping several pages of the directive. Condense most of your directives so that "Action" follows as soon after "Purpose" as possible.
   c. **Designate Responsibilities by Individual.** In the "Action" paragraph, itemize exactly what each individual must do. Readers will take notice if they see specific responsibilities assigned to them.
   d. **Use Inventive Paragraph Headings.** The only two paragraphs required for all directives are those for purpose and action. Standard paragraph headings like "Objectives," "Scope," and so on are not necessarily helpful. On the other hand, using more pointed titles (as illustrated here) can sometimes help the reader along.
   e. **Get Attention by Using Information Design.** Nothing in references (a), (b), or (c) prohibits you from using modern typographic techniques, especially bold face type, italics, indentation, and white space.

                                                                    A. T. Mahan

short-term information. Otherwise, they have the same force as Navy instructions and Marine Corps orders. Figure 3.12 is a sample Navy notice that discusses how to write clear and usable directives.

THE NAVY INSTRUCTION AND THE MARINE CORPS ORDER

Differing from notices and bulletins, instructions and orders are relatively long term. They have continuing reference value or require continuing action. Since these directives are relatively permanent (more often revised than canceled or superseded), they govern most major administrative efforts within naval commands. If we don't write them well, our units, programs, and communities will suffer.

FIGURE 3.13. *Format and Guidance for Writing Navy Instructions*

USS EXAMPLE INSTRUCTION 5215.2B

From: Commanding Officer, USS EXAMPLE

EXAMPLEINST 5215.2B
ADMIN: res
10 Apr 24

Subj :   WRITING THE "ACTION" PARAGRAPH AS A "TASKING" DEVICE

1. Purpose. To guide instruction writers in composing the "Action" sections of directives: the best course is to *allocate each task to a specific billet*.

2. Cancellation. EXAMPLEINST 5215.2A is hereby superseded.

3. Rationale. Specifying detailed assignments to individuals in a directive (as in the "Action" paragraph below) does several things:

    a. **It gets the attention of readers**. Sailors will learn to look for their responsibilities in the "Action" paragraph. A three-page instruction suddenly becomes readable if, in effect, all you have to read is the "Purpose" statement and one "Action" paragraph directed specifically at you.

    b. **It makes drafters do all the vital spade work**. Assigning specific tasking by billet forces a writer to articulate general principles into specific responsibilities. It helps our people think through and specify in full detail how policies or programs will be made to work.

4. Action. Take action as outlined below:

    a. The **Executive Officer** will ensure that all shipboard directives embody "tasking sections by billet," as appropriate to their content.

    b. **Department Heads** will supervise the training of their junior people in writing directives with good "tasking" paragraphs.

    c. The **Training Officer** will prepare lesson plans on directive writing, to feature "tasking" procedures prominently, and will furnish such materials to those conducting training.

    d. **All writers of directives** will specify in the "Action" paragraphs of their directives specifically who (by billet title) is responsible for exactly what and, if appropriate, when, where, why, and how.

                                                                    D. G. FARRAGUT

Distribution:
List 1, Case A

Again, as in the case of notices and bulletins, drafting a good "Action" paragraph is the central skill. "Discussion should be brief; action is the key; and action must be indicated by *job title* or *billet*," as one commander pointed out. Learn to draft brief and effective instructions or orders. Consider the sample Navy instruction in figure 3.13, which both demonstrates and explains how to assign responsibility in the format of a command instruction.

REVISING DIRECTIVES

Most of your work will be revising directives. Even if you must draft a completely new directive, you can often find an old one on a similar topic to guide you. As a Marine lieutenant colonel commented, "When you have to put out a new directive, find parallel orders and plagiarize from them, keeping the format." Similarly, a Navy commander suggested that you "plagiarize where you can. Someone else probably already wrote a similar directive. So go find it, change the names, and use it."

While altering an existing document for your purpose—changing, adding, or deleting details—also work to *revise for readability*. The principles for writing a directive clearly are similar to those of effective emails and other staff writing:

Adapt your content to your audience. Don't make it longer than it needs to be.

Make your subject line specific. "Onboarding Process for New Personnel" is more helpful than "New Personnel."

Put the action up front.

Address the reader with "you." As much as possible, talk directly to your audience.

Use active voice. Instead of "All leave must be submitted two weeks in advance. Each request will be submitted to the XO," write, "Submit your leave requests to the XO two weeks in advance."

Keep lists parallel. The rhythm of parallelism sets up expectations that make reading easy. A common violation of parallelism is switching from active to passive or from verb openings to nouns. Either is fine, but stay consistent with grammatical structures. The italic sentences that open the paragraphs in this list, for example, would be nonparallel if the third paragraph read "Action up front" or if this one read "Parallelism is important."

These steps will help ensure the directive you're revising will begin to govern action instead of just providing window dressing for an inspection.

~ 4 ~

Naval Messages

"Message traffic retains an important place—because it is unquestionably official."

—NAVY CHIEF

EXECUTIVE SUMMARY

Naval messages are perhaps the most distinct type of document prepared by Navy and Marine Corps personnel. The guidance for each section is much more specifically defined, and the appearance of messages is different from most others. Although email has assumed some of its functions, we still use the naval message for certain situations—frequently transmitted via email applications. Although naval messages appear different and require special arrangement and formatting, at heart the success of the main message still depends on tailoring your content, structure, and style for the audience, purpose, and situation.

NAVAL MESSAGE USAGE

"Write and read your messages as your seniors will usually read them—with no knowledge of the situation other than what you will tell them."

—ADMIRAL, SKIPPER OF A SURFACE ACTION GROUP

Naval messages—message traffic—is the most distinctive form of naval writing covered in this book, at least in appearance. OPNAVINST F3100.6J governs the use of messaging for special incident reporting. Although email provides a faster and clearer form of official written communication, naval commands still write and receive many naval messages, and certain occasions require the transmission of an official message. Often, message traffic is sent via email even though it follows official message format. Ships could not receive or respond to orders without them—quick reaction to emergencies throughout the world depends heavily on the naval message. Staff must often communicate with underway units on administrative as well as operational matters. Efficient coordination of repairs and the

---

FIGURE 4.1. *Naval Message Guidance*

R 151001Z JAN 23

FM   USNI GNW AUTHORS//NI//
TO   NAVAL PERSONNEL//JJJ//
            MARINE CORPS PERSONNEL//JJJ//

INFO   CIVILIAN DON PERSONNEL//JJJ//

UNCLAS //N01000//

MSGID/GENADMIN/TEXTAUTHOR NI//

SUBJ/PREPARATION OF STANDARD NAVAL MESSAGE//

RMKS/1. MSG CIRCUITS ARE OFTEN TIED UP, ESP DURING CRISES. IN HIGH TEMPO OPS OR DURING COMBAT, EVERY PRECEDENCE IS AT LEAST IMMEDIATE, AND FLASH MSGS CAN TAKE OVER AN HOUR TO TRANSMIT. MSGS ARE ALSO EXPENSIVE. LIMIT NAVAL MSGS TO URGENT COMMS THAT CANNOT REPEAT CANNOT BE HANDLED BY OTHER MEANS.

2.   MAKE USE OF EMAIL, FAX, PHONE, AND MAIL TO MAXIMUM EXTENT POSSIBLE, ESP WHEN COORDINATING W/STAFFS.

3.   LIMIT MSG SIZE. CUT OUT UNNEEDED WORDS. FREELY USE ABBREVS. IMAGINE EACH WORD COSTS A DOLLAR AND HONE TEXT. CUT PAGES, PARAS, SENTENCES, WORDS, EVEN LTRS.

4.   DON'T BURY ACTION. FIVE PAGE MSGS WITH ACTION AT END, THOUGH COMMON, ARE COUNTERPRODUCTIVE. PUT ACTION UP FRONT. USE SUBJ LINE AS TITLE, NOT JUST ROUTING DEVICE.

5.   PRACTICE ART OF MSG WRITING. WATCH HOW CO, XO, CSO WRITE MSGS, HOW THEY EDIT YOURS. NOTICE POLITICS OF MSGS, IMAGE PUT ACROSS, TONE, PROTOCOL, EFFECTIVENESS ABOVE ALL. LEARN TO GET THE MSG THRU.

BT

---

ordering of supplies, ordnance, or spare parts for vessels, embarked aviation squadrons, and embarked or deployed Marine units also must often go by naval message, though in recent years emails have taken over some of that coordinating and administrative load.

Message traffic is in one sense the most official and highly visible form of naval writing. Messages that take time to comprehend, that omit crucial information, or that violate protocol or procedure can easily sour the image of your command. Figure 4.1 offers an opening sample message that provides a summary of the principles of message writing.

The Naval Message, Section by Section

Writing messages well means much more than just the technical matters of getting all the characters in the right fields or blocks. Although such matters carry weight, even more weighty are such subjects as what gets said, to whom, from whom, and why. In other words, the content is much more important than the form, and you need first to understand the basic communication situation.

So, you need to know what each major section of a "GENADMIN" naval message provides. Appropriately, we begin with that key element we have discussed several times before—the audience. The first section of a naval message is the address.

THE ADDRESS: SEND IT TO THE RIGHT PEOPLE

Obviously, writing a top-notch and timely naval message does no good if you don't send it to the right people. Get a strong grasp of command relationships and address all who really need to know.

But who really does need to know? You'll quickly discover that there is much more than just technical accuracy to addressing messages—"political" aspects also play a role. You need to have a good grasp of how significant messages differ from naval letters in their routing and in who reads them.

Context: Situational Awareness

Realize who will read your message and why. To begin with, on board ship perhaps 90 percent of incoming messages are read by COs. Moreover, on a ship or operational staff, all officers and many chiefs will also read most incoming messages. (Sometimes these will be seen on an onboard message board as in the past; sometimes they will appear on Navy/Marine Corps Internet [NMCI].)

Beyond the attention of COs in the fleet or field, flag officers read messages avidly. If you accidentally add COMNAVAIRLANT to your message as an INFO addressee, NAVAIRLANT him- or herself (the admiral) will probably read your message. So when addressing your message, not only do you want to make sure that everyone who is supposed to see the message gets it and that you have listed all the required ACTION or INFO addressees, but you also must be sensitive about who (from your CO's point of view) should not see it. Young command duty officers have probably been called on the carpet more often because of to whom they addressed a message than for what they actually said in it.

One Navy ship nearing port in the northern Pacific began conducting a helicopter operation to get the mail off the ship expeditiously. In the process, a fifteen-pound bag was accidentally blown over the side and sank. The official responsible for the mail—the chief petty officer who was mail clerk—drafted the required message about the loss of the mail sack and included, as INFO addressees, CINCPACFLT, COMNAVAIRPAC, and the embarked flag. Soon the captain summoned him. "Who's required to get this message?" the captain asked. "The mail office," responded the chief. "How about these other addressees? Why are they here?" "Well, the flag had some mail in that bag," replied the chief. "How about the others?" "Just general information," said the chief. "Then don't send it to them," replied the captain.

The captain knew the admirals whose commands were listed as INFO address-ees in that message would probably read each one personally, and he made a policy of not reporting mishaps to anyone except those with a specific need to know. Obviously, we all have a moral and legal obligation to inform our seniors and those affected of significant mishaps. But this captain knew that "what people know of your command is what they read," and neither he nor any other alert CO wants every detail of the command's activities spread out unselectively in front of seniors.

Just as with a "cc:" in an email, sending a message INFO to an admiral is a com-mon way to try to get action taken, especially if the addressee responsible for action hasn't responded to the originator's problem. Know the consequences of addressing messages. As an LDO on a carrier commented, "Don't get the elephants involved by chance."

Remember Protocol

> On a former skipper's advice: "He told me: 'There is a tone that can be used in a
> message, a quality of junior talking to a senior, up the chain. Remember, we're
> the littlest frigate in the Navy, so don't forget to say please and thank you.' The
> difference was immediately evident in the way people responded to the ship."
> —LIEUTENANT COMMANDER

> "In messages we are SAMUEL GOMPERS; not SAM, FAT SAM, GOMPERS, the ship,
> the platform, or anything else. Other ships are also addressed by their full names."
> —GUIDANCE ISSUED ON USS SAMUEL GOMPERS (AD 37)

Do not neglect simple rules of naval courtesy. Separate action from information addressees first, then within either group, list addressees by proper protocol: highest echelons before lower, then by alphabetical order within echelons. Again, remember that all correspondence out of a command is a direct reflection on that command—this principle is especially true of messages, which have such poten-tially wide and senior audiences.

Of course, observe protocol not only in the address element but also throughout the message. Remember the assumptions that go with certain usages. One familiar piece of naval wisdom is, "Seniors *direct* attention while juniors *request or invite* attention to an issue or problem."

Be Consistent with Plain Language Addresses

The governing publication is the "Telecommunications Users Manual," Navy Telecommunication Procedure NTP-3. As this guide points out, the use of automated message-processing systems has made consistency in format and spelling of plain-language addresses (PLAs) critical. (This use of "plain language" is not the same as the general use of the term to mean clear, audience-friendly writing as discussed in chapter 1.) If you want the message to get to its destination on time, don't rely on memory. Instead, look up all military-wide standard addresses in the Message Address Directory; "USN PLAD 1" is the naval section of that publication. Here is further specific guidance:

Spell out numbers in PLAs:

· 1–19 as one word: EIGHT or ELEVEN or THIRTEEN, etc.

· 20 and up: COMDESRON FIVE ZERO or TASK FORCE NINE FIVE PT THREE, etc.

Spell out letter designations phonetically:

· FAIRECONRON ONE DET ALFA

Use no punctuation within each PLA:

· "PT" for period, "DASH" for hyphen, etc.

Specify Office Codes in the Addresses of Naval Shore Activities

"Make sure you use the right code when sending a message to a staff. If an office code in the addressees of a message is wrong, the message will go to the wrong guy, he'll ignore it, and all the work you put into crafting the RMKS will go for naught."

—NAVY CAPTAIN, OPNAV

Navy shore commands are often much bigger than ships, and their decision-making processes are much less centralized. So NTP-3 requires that we spell out office codes in the PLAs of all Navy shore activities. If you list more than one code at the end of any PLA, the first one listed should designate the one responsible for action.

Of course, familiarity with the ordinary responsibilities of such shipboard departments as CIC (combat information center), OPS (operations), ENG (engineering), and WEPS (weapons) does not necessarily help us understand what offices make decisions at a shore command. If you do not know the office responsible for the subject of your message, use the letters "JJJ" (following the PLA and enclosed by double slants) in place of the unknown code. Having used //JJJ// for any addressee once, however, take note of the code of the shore command that responds to your message and address that code in subsequent messages or other correspondence.

THE SUBJECT LINE: MAKE IT A TITLE, NOT A ROUTING DEVICE

As with an email, memo, or letter, the subject line of a message helps key the reader to its main issue, gets the reader's attention, and aids a reader in skimming the text for what is pertinent to that person's own area of concern. Composing a good subject line takes some care. Remember the *Correspondence Manual*'s excellent advice: Use the subject line to avoid mystery stories. Announcing the topic in the subject line prepares the reader for what is to come and helps get the right people to read the message.

> Instead of
> > Message Handling
>
> write
> > Recommended Changes in Squadron Message Handling

> Instead of
> > BAQ Entitlement
>
> write
> > Decision on BAQ Entitlement for Member Married to Member

If a message were titled just "Damage Control" rather than "Lessons Learned from Fighting Fire aboard the USS Cole," it would clearly attract fewer readers and maybe not the most important ones.

Descriptive titles will also help anyone who is searching through files to find the right message quickly—such titles also help those service members who have to route a message within a large command. Naturally, there are some subjects that don't require more than perfunctory titles. And in most cases, a subject longer than a line or two will just slow the reader down.

For a message about a man-overboard situation, "Man Overboard Lessons Learned" was a suitably brief subject line because everyone in the Submarine Force to whom it was directed would have already known what this referred to. The complete message first outlined the circumstances of the event and then proceeded to cite lessons that would have to be taken to heart so such a tragedy would not recur.

REFERENCES

With tightened security rules, sharply limited distribution, and requirements to destroy files earlier, depending on references is neither efficient nor dependable. Unless the action officer can lay hands on that original copy in the correspondence file, the reference simply may not be available. Under these circumstances, drafting messages that depend on information in prior message traffic is a sure recipe for delay, retransmittal requests, and misunderstandings. The lesson is to make your message stand alone as much as possible.

Referencing and Summarizing Prior Messages and Other Documents—The Pros

Should we try to avoid referring to prior messages and other references entirely? Yes and no. Referencing past messages or directives and summarizing their import in an "Amplification" (AMPN) line or short "Narrative" (NARR) paragraph (standard GENADMIN procedures) can sometimes serve important purposes.

If a message coming into a staff requires action from higher than the ordinary action officer, then in addition to preparing a recommended response, the action officer on the staff usually has to prepare a briefing memo for the admiral. A message that summarizes pertinent references cuts the time spent preparing the briefing memo.

Another reason for summarizing the background for the addressee's action officer or commander is that it allows readers to get a good feel for the context. This helps them see what you are asking or arguing and encourages them to proceed to an immediate decision. If you don't handily bring all relevant information to bear right there in the message itself but instead depend on the addressee to look up the references, the decision maker may want to look up all the past message traffic on the subject, read it all carefully, and think about it a while before proceeding to a decision.

Using this same line of reasoning, you can occasionally use a summary of the relevant past emails, messages, and directives to "prompt the witness." If your addressees have to go back and read several detailed references, they may come up with a different conclusion than you have (especially on a complex subject). But if you summarize the past references, you can make sure that nothing essential (from your point of view) is overlooked. Thus, your summaries can make it much more likely for the addressees to agree with you. For them to disagree formally, they have to go to the extra effort of looking up and reading all the references themselves.

Referencing Past Documents—The Cons

There's another side to this issue, however. For one thing, a commander giving an order seldom needs to cite all the references. Often, the commander simply says,

"Do it," and, except for making clear exactly what is wanted and why, has no need to refer to the past at all. The commander is establishing a new procedure, thereby cleaning the slate.

For another thing, both listing past references and summarizing them have often gone much too far. A summary has become for many message and email drafters a habit or crutch, clogging up message traffic with unnecessary volume and encouraging skimming. The summary must have a specific usefulness and a real need for the past references; using references is not just a drill to go through.

Clearly, much of the past referencing has not been helpful. Because of the requirement to refer to each reference somewhere in the message (originally an attempt to cut down on unnecessary references), we often find paragraphs like this one:

> REF A WAS RESPONDED TO BY REF B DELINEATING THE
> PROBLEMS WITH EXPANSION OF THE SUB-COST CENTER
> ORGANIZATION WITHIN THE G-3'S COST CENTER UNDER
> CURRENT DATA PROCESSING CONSTRAINTS. SUBSEQUENTLY,
> REF C WAS RECEIVED REQUESTING EXPANSION OF TRAINING AS
> A SEPARATE SUB-COST CENTER UNDER THE G-3. REF D RAISED
> THE SAME ISSUES AS REF A WHICH HAD BEEN ANSWERED BY
> REF B, AND REF E WAS FOLLOW-UP TO REF D.

Before writing such a paragraph, ask yourself: Do the addressees really need to know how negotiations have proceeded on this topic? If not, is there any other major reason for going into such detail? Often, the answer to both these questions is no.

Still, summarizing all the relevant correspondence on an issue may be important. For a senior command, each document may direct action on a certain aspect of the problem, and if so, all commands affected should have a complete file. Months may pass before the appropriate superior puts all those messages together into a comprehensive instruction.

But for many messages, just giving the general background is enough without extensive reference to past correspondence.

Referencing Past Documents: A Solution

Three conclusions seem reasonable—a three-step decision chain, as it were.

First, the general background is often enough by itself. Therefore, ask yourself whether the much clearer and more direct "CNO HAS REQUESTED" or "PRE-VIOUS COMMUNICATION HAS SPELLED OUT REQUIREMENTS FOR" can replace "REF A WAS RESPONDED TO BY REF B, WHICH CANCELED REF C," and so on. Notice the revision to active from passive voice here as well.

Second, if you believe a summary is either necessary or helpful, do your best to reference as few messages as possible and to make any summary as brief as you can. Strive above all to keep in mind the needs of the action officer you're addressing.

Third, in almost all cases, as with every other form of writing discussed in this book, put the main point up front in the very first paragraph. The GENADMIN format has made that possible in almost all cases.

THE TEXT: BOTTOM LINE UP FRONT

"It's traditional: in the first sentences of the first paragraph of a CASREP,

place the sentence to be read to the admiral."

—CHIEF ENGINEER ON A DESTROYER

Just as with so many other forms of communication, chief among concerns with writing the text is to get started right. If you do, the rest of the message tends to fall together nicely.

Jump Right in with the Action

How do you get to the main point quickly? Sometimes your addressees just need to know the action required or requested, with specifics in later paragraphs after a general announcement in the first. In this case an effective opening "Remarks" (RMKS) paragraph is simple and to the point. For example:

> RMKS/I. EFFECTIVE IMMEDIATELY DO NOT PROCESS PARTIAL
> PAYMENTS ON PREPAID RESUBMISSION INVOICES. PROCESS THE
> EXACT AMOUNT LISTED ON THE PITR.

This message opening informs the reader immediately of the basic action required; the reader may at leisure review the specific details found in subsequent paragraphs.

On other occasions an effective opening makes a simple, direct reference to a prior message that assigned a task or made a request:

> IAW REF A, FEEDBACK ON RELIABILITY OF ORDNANCE TEST
> EQUIPMENT FOLLOWS.

— or —

> THE USNTPS PREPARATORY CURRICULUM REQUESTED BY
> REF A IS APPROVED.

These openings are effective because they are clear and immediately grasped. The reader knows exactly what is to follow.

Use the BLUF and Impact Approach for Opening Summaries

Sometimes giving the reader a brief context for what follows is important. For this purpose, use a paragraph modeled after the BLUF approach.

For example, the following opening paragraph gives brief context and focus and then gets to the main point quickly, leaving details to follow in subsequent paragraphs:

> *Context*: 1. PLANS FOR CHANGE OF COMMAND ON 30 DECEMBER INCLUDE A RECEPTION FOR APPROX 700 IN HANGAR BAY OF USS IWO JIMA.
>
> *Impact*: AUGMENT OF SHIP'S FOOD SERVICE PERSONNEL IS NECESSARY TO ASSIST IN FOOD PREPARATION/SERVICE.
>
> *BLUF*: UNITS WILL RESPOND AS OUTLINED BELOW.

Such an opening prepares those with action obligations for the details that follow and gives a brief executive summary to others who can either stop there or read further for information. Whatever their needs, all readers have understood the gist of the message and have not had to wait several paragraphs to discover what on earth the writer is getting at.

Remember, except on long, detailed messages, you'll usually have no standard paragraph headings such as "Purpose," "Background," or "Action" that help the reader skim through the text. Briefing the whole matter in the first lines is as important in messages as it is in other kinds of naval writing.

A "Personal For" NAVADMIN That Quickly Gets to the Point

In a "Personal For" message, commanders speak personally to specific individuals or groups and back off from telegraphic style a bit to do so. Figure 4.2 offers a NAVADMIN, adapted from an actual message by VADM Frank L. Bowman many years ago, that quickly gets to the point, both in its subject line and in the very first sentences. It then provides telling details to drive the central ideas home to the audience.

STYLE FOR NAVAL MESSAGES

Use Telegraphic Style

The writing style for naval messages is similar to that of telegraphs. Many years ago, messages needed to be brief to save on transmission time over radio waves.

FIGURE 4.2. *NAVADMIN Example*

R 252128Z JUL 20

FM CNO WASHINGTON DC//N 1//

TO NAVADMIN

UNCLAS PERSONAL FOR COMMANDERS, COMMANDING OFFICERS AND
OFFICERS IN CHARGE//N00000//

NAVADMIN 134/20

MSGID/GENADMIN/PERS323//

SUBJ/TIMELY SUBMISSION OF PERFORMANCE EVALUATIONS AND FITNESS REPORTS//

RMKS/1. THE PURPOSE OF THIS NAVADMIN IS TO SOLICIT YOUR HELP ON AN ISSUE THAT HAS
A MAJOR IMPACT ON OUR MOST IMPORTANT RESOURCE, OUR PEOPLE. THE TIMELINESS AND
ACCURACY OF OFFICER FITNESS REPORTS AND ENLISTED EVALS ARE FAR TOO IMPORTANT FOR
YOU, THE LEADERS OF OUR NAVY, TO LET SLIDE. THE SINGLE MOST IMPORTANT FACTOR IN BOARD
DECISIONS IS PERFORMANCE AS DOCUMENTED IN EVALUATIONS AND FITREPS. AN INCOMPLETE
PERFORMANCE RECORD IS POTENTIALLY DETRIMENTAL TO A CANDIDATE'S CAREER. LAST YEAR,
OVER 100,000 EVALUATIONS WERE RECEIVED LATE, AND IN THE LAST THREE MONTHS ALONE,
6643 FITREPS WERE RECEIVED LATE. IN ADDITION, ALMOST 4000 MESSAGES REQUESTING MISSING
EVALUATIONS JUST FOR E7/E8/E9 SELECTION BOARDS WERE REQUIRED. SIMILAR REPORTING
DEFICIENCIES ARE NOTED BY EVERY OFFICER AND ENLISTED BOARD.

2. EVALUATION OF PERFORMANCE IS A FUNDAMENTAL RESPONSIBILITY OF COMMAND. FAILURE
   TO DO SO IN A TIMELY MANNER IS AN ABDICATION OF AN IMPORTANT RESPONSIBILITY.

3. REPORTING SENIORS MUST ENSURE THAT THE RECORDS OF ASSIGNED PERSONNEL ARE
   COMPLETE. THE ONLY WAY THE NAVY CAN BE SURE OUR PEOPLE'S ADVANCEMENT,
   ASSIGNMENT, AND TRAINING ARE HANDLED FAIRLY IS TO BE SURE THEIR FITREPS AND EVALS
   ARE CORRECTLY COMPLETED THE FIRST TIME AND ARE FORWARDED ON TIME. WE NEED YOUR
   HELP ON THIS.

4. RELEASED BY ALICE K. BROWN, VADM, USN.//

BT

This is much less of a concern now, but traditionally messages still aim for maximum brevity.

Do not waste words. Except in "Personal For" messages like the example in figure 4.2, you should normally leave out unnecessary adjectives, adverbs, prepositions, and most articles ("a," "an," "the").

Use truncated sentence structure and the imperative voice liberally. Instead of "we request that" say "request"; instead of "you should contact Mr. Herbert," say "contact Mr. Herbert."

Use small words instead of big ones where the sense is the same. Leave out words used only for the rhythm or aesthetic quality of a sentence.

The following paragraph from a Military Sealift Command letter can be revised into a brief and clear message style. The original letter, with wordy expressions italicized, reads:

> *In order* to eliminate *any* delay in *the* ammunition loading *operations, we* request *that, if possible, you* correct all deficiencies prior *to your ship's* arrival *at* Naval Weapons Station, Concord. *In the event* you do not have sufficient crew *members* on board *to accomplish this work, you are hereby authorized to* employ *the services of a* commercial contractor to correct *these* deficiencies.

Revised for standard message brevity:

> TO AVOID DELAY IN LOADING AMMO, REQUEST YOU CORRECT ALL DEFICIENCIES PRIOR ARRIVAL NWS CONCORD. IF TOO FEW CREW ON BOARD, YOU MAY HIRE COMMERCIAL CONTRACTOR TO CORRECT DEFICIENCIES.

Beware of False Economy

Though you should aim for concision, never sacrifice clarity for brevity. Remember, clarity is king.

Word a message so that, first and foremost, it clearly expresses the meaning you desire to convey.

Be sparing with punctuation, but use it where necessary for clarity or emphasis.

Though you may omit small prepositions and articles, don't omit little verbs such as "is," "are," or "was." These don't cost much in transmission time, and often their omission leads to confusion.

Abbreviate where useful with standard, well-understood abbreviations but don't overdo it. Sometimes words have to be spelled out or the transmission time saved will be much less than the time readers spend trying to decode or correctly construe unfamiliar shorthand.

Structure and Focus for Ease of Reading

Consider reworking longer messages so that, as in "bullet" format, a series of statements or questions falls under one governing statement. By using this method and by making a few other astute changes, a Marine colonel shortened the text of the following message (in pre-GENADMIN format) by one half. The message originally read:

FROM: USNA ANNAPOLIS MD
TO: CG FIRST MARBGDE HAWAII
BT
UNCLAS //1531//
SUBJ: MARINE SUMMER OPTION CRUISE
REF: A. PHONCON USNA MAX WHITE/FIRST MARBGDE CAPT PALANCIA OF 10 FEB 83

1. REF (A) REQ THAT ALL REQUESTS FOR INFORMATION CONCERNING THE MARINE OPTION CRUISE BE SUB BY MSG.

2. IAW REF A, THE FOLLOWING INFO REQ.

   A. IT IS DESIRED THAT THE MIDSHIPMEN BE ALLOWED TO EAT IN THE ENLISTED DINING FACILITY: WHAT PAPERWORK IS NECESSARY?

   B. WHAT IS THE TOTAL NUMBER OF MIDSHIPMEN THAT THE BRIGADE CAN HANDLE? IS ADEQUATE BOQ SPACE AVAILABLE FOR THIS NUMBER?

   C. DURING THE SUMMER OF 1982, A SCUBA TRAINING PROGRAM WAS OFFERED, IN THE LATE AFTERNOON. CAN THE PROGRAM BE OFFERED IN 1983?

   D. IT IS NOT DESIRED (FOR FINANCIAL REASONS) THAT THE MIDSHIPMEN MISS ANY MEALS. CAN "C" RATIONS BE PURCHASED THROUGH THE BRIGADE TO PREVENT MISSED MEALS?

   E. IS IT POSSIBLE TO PROVIDE A STANDARD WELCOME ABOARD PACKAGE FOR EACH MIDSHIPMAN WHO PARTICIPATES IN THE PROGRAM?

   F. CAN THE MIDSHIPMEN GET THEIR AVIATION PHYSIOLOGY TESTING COMPLETED IN HAWAII?

   G. SEVERAL OF THE MIDSHIPMEN ARE AIRBORNE QUALIFIED. WILL THERE BE AN OPPORTUNITY FOR THEM TO JUMP WHILE THEY ARE WITH THE BRIGADE?

3. MAJ WHITE IS CURRENTLY SCHEDULED TO ARRIVE IN HAWAII ON 4 MARCH. HE WILL BE AVAILABLE UNTIL 11 MARCH TO ACCOMPLISH ANY COORDINATION NECESSARY FOR THIS PROGRAM. CAN HE MEET WITH THE BRIGADE LIAISON OFFICER DURING THIS TIME?

The revised message reads as follows:

FM USNA ANNAPOLIS MD
TO CG FIRST MARBGDE HAWAII
BT
UNCLAS //N01531//
SUBJ: MARINE SUMMER OPTION CRUISE
1. MAJ WHITE, USNA LIAISON OFCR FOR MARINE SUMMER
    OPTION CRUISE, WILL BE IN HAWAII 4-11 MAR TO
    COORDINATE WITH 1ST BDE REGARDING THIS PROGRAM.
    REQUEST HE MEET WITH BDE LIAISON OFCR AT THIS TIME TO
    DETERMINE IF MIDSHIPMEN CAN:
    A. BE BILLETED IN BOQ? HOW MANY?
    B. EAT IN ENLISTED MESS?
    C. BE FED ALL MEALS?(C-RATIONS ACCEPTABLE WHEN
        ENLISTED MESS NAVL; USNA WILL PURCHASE IF REQUIRED)
    D. PARTICIPATE IN SCUBA TRAINING?
    E. BE PROVIDED STANDARD WELCOME ABOARD PACKAGE?
    F. GET AVIATION PHYSIOLOGY TESTING COMPLETED?
    G. PARTICIPATE IN AIRBORNE JUMPS? (QUALIFIED
        MIDSHIPMEN ONLY)
    BT

The major changes in the revision are as follows:

- Eliminated unneeded "Reference A" and all subsequent references to it
- Combined separate questions in lines A–G into one question with seven parts
- Omitted reasons and background wherever obvious or unnecessary
- Inventively reduced wording throughout; used parallel structure for all verbs to fit with overall question "If midshipmen can . . ."
- Condensed 3 paragraphs into 1

SPECIFY THE FORMAT FOR THE REPLY IN TASKING MESSAGES

Tasking people to do work for you is an art. A DESRON staff officer outlined a technique for tasking ships to send information to the squadron. He said that to get good input, you should be sure to specify the following:

- The titles of applicable references—and sometimes you have to actually send the references too.

- A date to respond by, the date selected so you can collate all information and draft a response to your superior by your due date.

- A format that helps you work with the data—that is, a format specifying both what data you want and in what order.

- Specific units of measurement required; if you don't specify, they'll send you a general answer, then you'll have to go back again to ask for more.

- A point of contact at your location, including a phone number and email address.

And as the following passage shows, sometimes ambiguity in a naval message can be used to advantage:

In the early '60s, a Captain led his Mediterranean Destroyer Flotilla into the Black Sea, where he soon sighted a squadron of Russian Cruisers closing his flotilla at high speed.

From leading Russian Cruiser (by light):

WHAT ARE YOU DOING IN THE BLACK SEA.

Turmoil ensued amongst the Staff on the flotilla-leader's bridge while signal logs were sent for and Diplomatic Clearance discussed. At last the Captain raised an elegant hand for silence, and said quietly to the Signalman,

Reply:

TWENTY-ONE KNOTS.

— Captain Jack Broome, *Make Another Signal*
(London: William Kimber, 1973), 248.

A LAST WORD

Besides learning prose-cutting techniques, abbreviations, and so on, there is at least one other method of keeping costs down and the circuits clear. Limit all electrical transmissions to urgent official business that other means cannot satisfactorily handle. Use email or telephone whenever possible in place of naval messages, reserving this form of writing for only the most important situations.

FIGURE 4.3. *NAVADMIN Message Using Mixed Case, Not All Caps*

UNCLASSIFIED//
ROUTINE
R 292122Z NOV 21 MID600051229095U
FM CNO WASHINGTON DC
TO NAVADMIN
INFO CNO WASHINGTON DC
BT
UNCLAS

NAVADMIN 272/21

PASS TO OFFICE CODES:
FM CNO WASHINGTON DC//N1//
INFO CNO WASHINGTON DC//N1//
MSGID/GENADMIN/CNO WASHINGTON DC/N1/NOV//

SUBJ/PUBLICATION OF BUPERSINST 1610.10F (EVALMAN)//

REF/A/DOC/BUPERS/06DEC19//
REF/B/NAVADMIN/CNO WASHINGTON DC/201449ZDEC18//
NARR/REF A IS BUPERSINST 1610.10E, NAVY PERFORMANCE EVALUATION SYSTEM. REF B IS NAVADMIN
312/18, ADVANCEMENT POLICY UPDATE.//

RMKS/1. This NAVADMIN announces the cancellation of reference (a) and the publication of
BUPERSINST 1610.10F, Navy Performance Evaluation System (EVALMAN).

2. Summary of Changes: This revision incorporates policy guidance contained in reference (b). The
    following new guidance applies with the updated instruction:
    a. Rescind a requirement for administrative change requests to be submitted within two years of the
       performance evaluation end date.
    b. Incorporates multiple changes as part of the introduction of eNavFit. The eNavFit interface
       consolidates the functionality of the current NAVFIT98A application with the Navy Personnel
       Command (PERS-32) processing capabilities into a uniform solution for the Navy. This interface is a
       functionality bridge and is not meant to replace or modify the performance evaluation process. The
       online and web-based solution allows users to create, edit, delete, route, validate and digitally sign
       reports. It also allows a reporting senior to group and process summary group reports. The online
       application is capable of receiving performance evaluation data from the offline version and will
       be deployed and operated within BUPERS Online (BOL) as the front end for Fleet users. The offline
       version uses Adobe Reader forms and provides many of the same capabilities as the online version.
       (1) Online performance evaluations submitted to BOL shall be digitally signed. The eNavFit user
           guide can be reviewed and downloaded at: https://www.mynavyhr.navy.mil/Career
           -Management/Performance-Evaluation/.
       (2) Offline performance appraisals submitted via mail (FEDEX, USPS, UPS, DHL, etc.) shall be wet
           signed using only black or blue-black ink.
       (3) Summary letters are not required for electronic submissions via BOL.
       (4) Personal data may be auto populated from the BOL server. It should always be verified for correctness.
    c. Incorporates changes in chapter 18 for mid-term counseling and coaching integration. Conducting
       mid-term counseling is not optional.
       (1) Introduces the term supervisor as coach to replace the previous terminology of counselor due to
           the guidance, support and direction they provide during the performance counseling session.
       (2) Introduces the term Sailor as performer to replace the previous terminology of member due to
           the member providing honest and candid feedback regarding gaps and opportunities for them
           to enhance their performance.
       (3) Authorizes the individual development plan (IDP) to be used as an optional developmental tool
           and to serve as a performance counseling record and growth plan.
       (4) Introduces addressing specific areas of performance improvement by demonstrating genuine
           curiosity and asking questions to guide the member to create an actionable plan.

FIGURE 4.3. *(continued)*

(5) Incorporates changes to the supervisors preparation requirements to include reviewing the members self-appraisal, determine any opportunities for development and to create a rough outline of meeting talking points and goals to discuss.

(6) Incorporates changes to the members preparation requirements to include preparing a self-appraisal and preparing to receive both positive and constructive feedback.

(7) Introduces three core skills to be used during a meaningful performance counseling session to include active listening, empathy and asking powerful questions.

(8) Incorporates changes to the Guidelines for Supervisors to include the addition of five new steps titled: The Introduction, Engage the Member, Diving Deeper, Give Specific Feedback, and Ending the Performance Counseling Session.

(9) Incorporated changes to the follow-up/monitoring of the members performance counseling session to include the supervisor and member will agree to continue open and informal performance conversation after the mid-term performance counseling session has concluded.

d. Incorporates changes to regular reporting senior for Navy Reserve Unit Personnel.

(1) All members of Navy Reserve Units (e.g., commissioned/operational and readiness/augment) assigned to mobilization billets will be reported on by the commanding officer (CO) or officer in charge (OIC) of the unit mobilization unit identification code (UMUIC) effective 2 August 2022. Reports with a reporting senior signature from either the UMUIC or training unit identification code (TRUIC) will be accepted to support a transition period of 1 February 2022 to 1 August 2022.

(2) UMUIC and TRUIC leadership will coordinate during the transition period to determine appropriate submissions/content, taking into consideration professional growth, development, continuity and member advocacy to minimize any/all disruptions related to officer fitness reports, enlisted evaluations and Navy wide advancement exam cycles. Members who are not assigned to a mobilization billet (e.g. who are in an in an assignment processing (IAP) status) do not have an associated UMUIC and therefore the reporting senior will be the TRUIC CO/OIC.

(3) The reporting senior for IAP members in readiness support units (RSU), formerly known as operational support units (OSU), and those in voluntary training units (VTU), will be the designated TRUIC CO/OIC or the Navy operational support center CO.

e. Incorporates changes to concurrent reporting senior for Navy Reserve unit personnel. TRUIC reporting seniors shall follow procedures in chapter 4 to submit inactive duty concurrent reports on their shared cross-assigned members to recognize those who significantly contribute to their TRUIC billets via inactive duty for training. There will be a transition period of 1 February 2022 to 1 August 2022 and UMUIC/TRUIC leadership will coordinate to determine the most appropriate reporting senior for concurrent reports during the transition period.

f. Incorporates changes clarifying the use and what can be included in the comments section for the *N* code, Block 20, Physical Fitness Assessment Code, for pregnant service members.

(1) If using the *N* code because of pregnancy, no directed comment should be used in the performance comment section.

(2) Do not quote from medical reports or summaries and do not include comments pertaining to medical issues (physical and/or psychological, e.g. pregnancy, post-partum, etc.) that do not affect the members performance of duties and/or his or her effectiveness as a leader.

g. Incorporates reference (b) announcing reporting seniors shall incorporate their post summary group reporting seniors cumulative average (RSCA) score for E-5 and E-6 evaluations into Block 43 (Comments on Performance). This RSCA score will be on the last line of the comments.

3. For questions concerning these policy changes contact the MyNavy Career Center at (833) 833-6622 or via email at askmncc(at)navy.mil.

4. Released by Vice Admiral John B. Nowell, Jr, N1.//

BT
#0001
NNNN
UNCLASSIFIED//

Although the naval message has a long history of appearing in ALL CAPS, this requirement no longer applies. Figure 4.3 is a NAVADMIN message announcing an update to the performance management system in 2021 (with some relevance to the content of chapter 5). Although the special formatting at the top of the message remains the same, the primary content appears in more readable mixed-case type.

CHECKLIST FOR COMPOSING STANDARD NAVAL MESSAGES

Communication centers have checklists to help you double-check the technical details of message writing and message handling. This checklist has a different intent: It is formulated to help you get the right points across.

- ✓ Have you included all necessary ACTION addressees? Included all INFO addressees?
- ✓ Should you omit any addressees?
- ✓ Within respective categories, are addressees listed according to protocol?
- ✓ Have you included office codes for any Navy shore activity PLAs?
- ✓ Have you included the pertinent references?
- ✓ Can you do without any references?
- ✓ Is there any correspondence needed for understanding this message that addressees do not hold? If so, have you summarized it or sent it along?
- ✓ Will the subject line adequately identify the subject to all readers?
- ✓ Have you assigned the appropriate precedence?
- ✓ Is the classification proper, and have you followed all proper procedures for classification?
- ✓ Does the main point appear in the first paragraph of the message text?
- ✓ Consider the commands and officials receiving this message—will they all understand it? Do you need to add or clarify any statements?
- ✓ Can you eliminate explanations nobody needs?
- ✓ Consider the words carefully; if each one cost you a dollar, would you include all those you have written?
- ✓ Could you condense the message substantially by changing the format?
- ✓ On the other hand, have you unwittingly stressed brevity over clarity? Is the message understandable? Are all abbreviations standard and clear?
- ✓ Is the paragraphing logical?
- ✓ Can you use indentation or headings to good effect?
- ✓ Will anything in this particular message or the way it is written give the command a bad image?

✓ Have you included all of the essential—but only the essential—
information? Has everyone seen this message who should see it before it
goes out?

✓ Does this information really have to go by message?

✓ Will the CO release this message?

CASREP MESSAGES

Besides standard naval messages in GENADMIN or other specified formats, naval
communicators use a variety of fully "formatted" messages designed to ensure that
all specific kinds of crucial information—ship's movement, logistic requirements,
and others—gets from a ship to multiple audiences regularly and efficiently.

By filing a MOVREP, for example, a communicator won't have to reinvent the
wheel when announcing the ship is getting underway—something each vessel does
dozens of times during a year. The addressees are standardized so the communicator won't forget anyone who ought to know this information. Typically, this
message must go to a great many addressees.

You might think that writing formatted messages would just be filling in the
blanks. But mere bits of information can never communicate everything we mean,
so throughout formatted messages, we find AMPN lines and NARR and RMKS
sections, each of which calls for clear, brief naval prose. Take, for example, one of
the most critical of all naval messages in the surface Navy, a form also important to
submarines and aviation units: the casualty report (CASREP).

WRITING CASREP REMARKS

> *"Make sure you comment on the impact of the casualty on the command*
> *schedule—what you won't be able to do next week, what problems the casualty*
> *will present in the ship's sked. The admiral himself won't pick up the impact*
> *necessarily just by being told the specific casualty; the staff might not either; but*
> *the ship will certainly know. And the admiral goes up the wall if not told."*
>
> —COMMANDER, XO OF A DD

You need a remarks section in a CASREP because no amount of formatted information can convey the exact technical information required to fix the casualty.
Especially on surface ships, the exact consequences of a specific casualty on a ship's
operations and the difficulty of fixing the problem will vary widely from vessel to
vessel, even among identical ship types.

Leading petty officers, chiefs, and officers at all levels may have to write or revise CASREPs. On some ships division officers write the whole report even though they often don't have specific technical expertise on the particular equipment involved. No matter who does the basic draft, COs and XOs give intense scrutiny to the remarks section (checking for pertinence, completeness, and brevity) before a CASREP leaves the ship. For many reasons, writing a CASREP is a high-priority item at most commands.

Write Your CASREP Remarks for Multiple Audiences

Who will read your CASREP? A great number of people at many commands. Some of them will be experts in the particular technical area involved, but most of them probably will not be. Your RMKS section must serve both expert and nonexpert audiences. Solve this dilemma by following three procedures. First, use headings and indentations to help the reader. Second, pen an executive summary at the beginning of the section for your operational commanders and other "generalists." Third, follow this summary with a detailed description composed for the technical expert.

Use Headings and Indentations

Just as they help the reader navigate any notice, instruction, or technical report, typographical features can help one find crucial information in a glance at a CASREP. NTP-3 allows indentation in naval messages when it will increase readability or clarity. The free-form nature of RMKS and AMPN sections also allows for the use of headings.

Write an Executive Summary for Commanders and Nonexperts

Suppose you serve as an officer aboard a combatant and it suffers damage. Your flag officer and other superiors will want to know the ship's exact condition. They need to know what it is still capable of, when it will be ready, where it will go for repairs, and how to expedite the repair process.

A ship's operational commander will tend to focus on the following areas of the CASREP, skimming the rest:

- the name of the ship
- the specific equipment reported—all in the formatted portion
- the C-rating
- the summary portion of the remarks

Obviously, your summary is important. A tested procedure must begin by carefully describing the exact nature and extent of the casualty. Write this description for a general Navy reader, not a technical expert. After this description, discuss the casualty's effect on the ship's immediate and near-future operational, exercise, or inspection schedule. Then speak to the mission degradation—that is, its effect on the ship's ability to carry out primary and secondary missions.

Operational and administrative commanders may have issued specific requirements (C-ratings and such) for what you must cover in these areas. If not, remembering to write for the informational needs of commanders and keeping your remarks general enough so that all of them can get the gist will allow both captain and chief to grasp the big picture immediately.

Follow the Summary with a Technical Description for the Experts

The technical description must supply the technical staff on the beach with enough detailed information so they can start the repair process in motion, sometimes well before the ship returns to port.

The ultimate readers will be personnel with technical expertise on the specific equipment reported. These experts will usually work at a repair facility and will arrange for parts, organize technical assistance, and act as the control point for all services. They will need detailed and specific technical information with which to begin.

Do not be obscure in this description. A duty officer or chief on the squadron or group staff must often get the repair process started, and neither of these people will necessarily be an expert in the specific area of the casualty. For example, the chief who reads the CASREP may be a chief electronics technician, whereas the casualty may be a piece of engineering equipment he or she has never seen. Nevertheless, this reader may have to interpret the message and initiate action in response to the CASREP without expert assistance, especially if the message comes in on a holiday or a weekend.

So, although the technical description should detail all the vital specifics, it should include as little jargon and technical obscurity as possible.

Be Careful about the Details

A chief on a naval supply ship learned just how vital specific details could be. Deciding he was obliged to do a CASREP, he looked at the specific piece of equipment in front of him, called it a "motor generator" on the CASREP form (the name everyone on board customarily gave the equipment), and reported the number he found in front of him.

The Navy Supply System, however, called only the bottom part of the equipment a "motor generator"; the equipment that fitted on the top had another name and a different stock number. It would be like requiring a lamp, for repair purposes, to be broken down into a lamp and a lamp shade. As a result, when the contractor assigned to the repair saw what needed attention, he said, "We can't go further with the repair; we haven't been contracted to work on the top part." The confusion cost the Navy an extra $2,000 and several days to straighten out.

Give the Repair Facility Enough Information

"In the Navy, it's important to be as self-sufficient as possible. So in point papers, emails, messages, or CASREPs, always include what you've done about the problem. Don't raise any new issue without answering this. Your boss or a shore command is much more likely to help if you've already started on the problem."

—LIEUTENANT COMMANDER ON A DESRON STAFF

If a ship is in its home port, completeness may not be crucial; a phone call or email can clear up questions pretty quickly. But if the vessel is underway or at another port, the facility vitally needs complete information. Any need to query you because you didn't put in enough detail could result in a substantial delay in the repair—as much as two to four days, according to a median estimate.

The Navy's general need to reduce fleet message traffic sometimes conflicts with the needs of the shore-based technical staff to get detailed answers for fixing vital equipment. Only experience will tell you how much is too much information.

One lieutenant wondered why the shore facility was advising him by message how to troubleshoot a type of equipment having a different voltage than the reported equipment on board his ship. Only later did he find out that the XO had cut out the specific voltage he had listed in the CASREP to shorten the message's length before it left the ship.

Think twice before cutting out details.

POOR CASREP REMARKS

Figure 4.4 contains the RMKS section of a CASREP from a ship away from its home port. The duty officer on the staff who received the report says that as soon as he read it, he knew SURFLANT would call. Sure enough, they did. But before that, the chief staff officer had already hit the roof. Why?

We've italicized particularly weak phrasing to illustrate some of the problems here. We don't know from this description what happened or how it happened. Did the technician break it? Is it completely down? Was it bad maintenance or a failure to follow proper maintenance? We can assume some things, but the writer doesn't

---

FIGURE 4.4. *CASREP Remarks Section with Insufficient Detail*

RMKS/MK 42 MOD 9 5" 54 ELECTRICAL SYSTEM *NOT FUNCTIONING*. NO IMPACT ON CURRENT OPERATIONS. *REDUCES SHIP'S AAW AND ASUW CAPABILITIES*. CASUALTY TO ELECTRICALLY DAMAGED SOLENOIDS THAT CONTROL CARRIER AND LOADER DRUM VALVE BLOCK ASSEMBLIES. SOLENOIDS ON ORDER PRIOR TO CASREP CONDITION. SHIP'S SKED: ENRT NEWPORT RI 23 JAN; IPT NEWPORT 25 JAN-26 JAN; IPT PHILADELPHIA 29 JAN-15 FEB.//

---

provide the oversight (SURFLANT) enough information for operational purposes or for repair.

GOOD CASREP REMARKS

Figure 4.5 is the RMKS section of a different CASREP, this one cited by the same duty officer as an excellent message. It affords a clear and concise summary, vital operational information, and a fully detailed technical description of the problem.

Here, we've italicized a few places that provide particularly helpful clarity. Notice how this example identifies both the specific casualty and the ship's continuing capability, thus clarifying the exact scope of the casualty. Identifying particular components as it does, this technical description describes the problem in detail. As a result, the repair agency may be able to troubleshoot the problem by message. This tells the staff that the problem is beyond the scope of Shore Intermediate Maintenance Activity (SIMA). It also shows that the ship is doing its best to be on top of things, already getting what help it can, and provides additional detail that may help the repair agency identify the problem.

FIGURE 4.5. *CASREP Remarks Section with Good Detail (Indentations Added)*

RMKS/CASUALTY: MK 53 ATTACK. CONSOLE WILL NOT POSITION KEEP THEREFORE WILL NOT
    UPDATE—PREDICT TARGET COURSE AND SPEED. MK 53 AC WILL ACCEPT NDT FROM ALL SONAR
    UNITS. MANUAL INPUT OF TARGET COURSE AND SPEED CAN BE USED FOR ASROC-TORP FIRINGS.
IMPACT: MAJOR IMPACT ON INSURV TO BE CONDUCTED WEEK OF 29 FEBRUARY.
MISSION DEGRADATION: MINOR IMPACT ON ASW MISSION AREA.
TECHNICAL DESCRIPTION: 12A1A4A15 MG2 (MOTOR-GENERATOR) HAS SAT INPUT BUT HAS
    INSUFFICIENT TORQUE TO TURN CLUTCH CM3 WHEN CLUTCH IS ENERGIZED. STRAY VOLTAGES
    HAVE BEEN INTERMITTENT AND BELIEVED TO BE RELATED TO CAUSE OF CASUALTY. WHEN
    "GOING INTO CONTACT" OR WHEN NDT FROM SONAR THE CLUTCH ENGAGES MOTOR BUT
    IMMEDIATELY SEIZES DUE TO INSUFFICIENT TORQUE STATED ABOVE, THEN BY PULLING
    RELAY KI OR "COMING OUT OF CONTACT" THE CLUTCH DISENGAGES AND SYSTEM WILL PK
    UNTIL NEXT NDT OR WHEN GOING BACK INTO CONTACT. SIMA NEWPORT R-5 ASSISTED IN
    TROUBLESHOOTING. CASUALTY DISCOVERED WHILE TROUBLESHOOTING TIMING CIRCUITRY
    FOR CONSTANT 7 SECOND FLASH. REQ TECH ASSIST TO DETERMINE WHERE STRAY VOLTAGES
    ARE COMING FROM.
SHIP'S SCHEDULE: 25 FEB ISE (PREINSURV); 25 FEB-7 MAR INPORT.

5

Performance Evaluations

"The focus in evals should be:

- *What did you do?*
- *What value did it have?*
- *What was the direct result?"*

—ASSISTANT MASTER CHIEF PETTY OFFICER OF THE NAVY

EXECUTIVE SUMMARY

The Navy and Marine Corps systems for evaluating performance and preserving a record of that assessment serves the primary purpose of helping selection boards choose the best enlisted Sailors and officers to promote. In a small amount of text, the writer of an evaluation (eval) needs to capture both a convincing overall assessment of potential and a clear, detailed description of the member's most impressive accomplishments. Thoughtful organization, concise statements, and careful word choice can make a tremendous difference in the effect the evaluation has on helping or hindering promotion.

THE PERFORMANCE EVALUATION SYSTEM

Patterned on the best of modern personnel systems, the Navy performance evaluation system, when used as designed, provides a powerful means of selecting the best possible people to be promoted and counseling them in their performance.

The system is fully spelled out in BUPERSINST 1610.10 (current edition). What follows here is guidance based on interviews done throughout the naval service; the many examples provided in this chapter are based on actual Navy evals and fitreps. Note that although counseling or coaching is a major part of the evaluation system, this chapter deals only with the writing of evaluations, not with the counseling. See the memorandum in figure 2.12 for a good example of a letter of instruction.

The Marine Corps fitness report system is explained in MCO 1610 (series). Most of the general writing guidance in this chapter is relevant to describing Marines' performance as it is to Navy personnel. But the Corps' system does include some important differences of context that writers need to understand. We discuss these below.

AUDIENCE, PURPOSE, AND CONTEXT FOR EVALUATIONS

> *"In theory, evals constitute an appraisal of performance. In reality, they are used*
>
> *by boards to help them render decisions on advancement. If you use an eval or*
>
> *fitrep to give someone feedback on his performance, you're doing him a disservice."*
>
> —CO OF A TRAINING COMMAND

PRIMARY AUDIENCE: THE SELECTION BOARD

Although several audiences read Navy enlisted evals and fitreps, you should write only for the most important of them: the formal Navy or Marine Corps selection board.

To properly prepare an evaluation for review by the board, you need to understand the perspective and reading context of its members. They have many records to review in a short time period. This longstanding practice has become more streamlined with new technology, but ultimately, board members cannot dwell overly long on any one evaluation. The leadership, technical skill, effectiveness, and relative strengths of the individual need to stand out clearly upon first reading it.

Enlisted boards typically review only the most recent evals in a service record, although officer briefers will usually examine all of the service member's fitreps. For example, the record of a commander going up for promotion to captain may include thirty or more fitreps. That's an enormous amount of reading for any particular board member, having many other commander's records to consider at the same time. Write the fitreps so that the briefers can easily find and highlight key points when they update the board members. We will discuss later some types of statements you should include and how to organize them.

Who's on the Board?

The breadth of a board's composition should also influence the content and structure of the evaluation comments. Although varying in composition depending on the level (mainly E-8s and E-9s on an E-7 board, Navy O-6s on an O-5 board, and so on), a selection board for line officers and chiefs is typically a mixed group. A line officer's board, for instance, comprises representatives from all line communities (for example, aviation, surface, and submarine). A chief's board is similar: many different rates are represented. Even though it may separate into smaller panels of similar ratings for some deliberations, any one of those panels may not contain two members of the same rate. Even on a restricted-line board, such as for the Supply Corps, which does comprise officers from that community, you can't expect each member to know the finer points or technical jargon of any particular job.

What Is the Board Looking For?

"We're looking for the whole YN. We like to see a person's work on the DC party, or work on PQS. And be specific, not generic: not 'She compiled 20 hours of community service,' but instead, 'She put in 20 hours with Habitat for Humanity.'"

—MASTER CHIEF YEOMAN

Selection board members are looking for the best and most highly qualified officers and enlisted personnel to promote. In determining "best and fully qualified," they must consider a great deal of specific information, including what a person has accomplished, what qualifications another person has achieved, and how the Navy or Marine Corps has profited from his or her performance. They must determine whether the write-up captures qualities important in a chief or officer. Besides demanding leadership and supervisory abilities, a chief's board will also diligently look for key qualities such as technical expertise, professionalism, loyalty, or active communication. Specific accomplishments can suggest that the candidate has some of these qualities, but you must also speak explicitly to the qualities themselves.

THE INTERNAL BOARD AT YOUR OWN COMMAND

As discussed above, the official Navy selection board is the primary decision-making audience for an eval or fitrep. Yet many commands hold a less formal but sometimes even more critical ranking board much closer to home, one that determines which officer gets the "must promote" designation or which chief will be marked in the "early promote" block. These decisions are typically not made by the XO or CO alone, especially in large commands, but instead by the carefully selected members of an internal ranking board.

There are no formalized rules about this process. Each command operates differently, but all must make such decisions. Most convene a group of chiefs or officers to assist in this.

Of course, the CO can approve or modify that ranking. But the point here is that even the draft evaluation, the informal and unsigned write-up, can affect the assignment of promotion recommendations (and perhaps even the trait markings) themselves. As one commander commented: "On a large shore staff, the write-ups can make a huge difference. The write-up may be the only thumbnail sketch of the chief he has at hand to inform him about the individual." The write-up becomes critical to the local decision.

The smaller the command, the more informal this internal ranking board becomes. Some officers have described the determinations as oral, with no one bringing the evaluation drafts with them. Nevertheless, even here the write-ups might affect the local process. All evals or fitreps must be signed by seniors, who upon seeing how much weaker one write-up is than another, may have the evaluations rewritten or may even revisit the recommendations.

Writing Effective Evaluations and Fitness Reports

"Remember that you are trying in most cases to write for impact—

impact on the CPO selection board, for instance. FACTS, FACTS, FACTS!

Navy guys are tough, and too much 'flowery' language conveys the idea that there

is no hardness to the ratee."

—NAVY CAPTAIN, FORMER HEAD, NAVY E-8/E-9 SELECTION BOARD

MARINE CORPS FITNESS REPORT COMMENTS IN CONTEXT

The Marine Corps uses the term "fitness report" for both enlisted and officer assessments. As we have explained in every chapter, you must understand the context as well as the audience of a piece of writing and make decisions about content, structure, and style. As required in MC 1610 (series), the Marine Corps form and approach focus heavily on specific qualities. The reporting senior (RS) assigns scores, ranging from A to G (G being the very best), to the Marine reported on (MRO) in thirteen different attributes. These convert to a numerical average, although this raw average is unimportant. What matters is how that average score compares with the average scores the same RS has historically given others of the same rank.

In addition, Marine Corps fitness reports usually group service members of the same rank within a unit into a top third, middle third, and bottom third. The form also includes a visual chart (sometimes called a "Christmas tree") for identifying how the MRO compares to others in the same grading category.

Within this context, the RS must draft written comments describing the most noteworthy qualities and accomplishments (or the absence thereof) in content and

> **FIGURE 5.1.** *Examples of Marine Corps Fitness Report Comments*
>
> **Top Third**
> Sergeant Washington perfectly models the professionalism, loyalty to the mission, leadership, and technical skill desired in all Marines. He and his staff managed and maintained the unit's weapons flawlessly, ensuring zero down time in operational exercises. He trained his Marines in weapons maintenance and handling so effectively that 100% of them received commendation letters and top marks on the unit weapons inspection. Sergeant Washington also possesses the highest moral standards and demonstrates the utmost integrity for his troops and those above him. An absolute must for promotion and assignments of greater responsibility.
>
> **Middle Third**
> Sergeant Washington has shown superior professionalism, leadership, and technical skill in his MOS. He expertly maintained the unit's weapons with minimal down time, ensuring strong support for operational exercises. A mature leader, he worked closely with his Marines in weapons maintenance and handling, ensuring their technical competence as well. Sergeant Washington also possesses strong character and supports the chain of command. Strongly recommended for promotion.
>
> **Bottom Third**
> Sergeant Washington demonstrates the technical skill and professionalism expected for his position. He maintained the unit's weapons effectively, fulfilling all requirements from his superiors. As a leader, he trained his Marines in the basics of weapons maintenance and handling with no incidents. Promote.

style consistent with the attribute grades and the ranking group assigned. Written comments that say the Marine excels in every endeavor and always achieves superlative results alongside a mediocre set of numerical grades will undermine the credibility of both the grades and the written comments.

Written comments need to address three elements: proficiency in technical skills, leadership, and an endorsement for promotion and added scope of responsibility. In addition to those written by the RS, the reporting officer (RO)—to whom the RS reports—may add comments that, ideally, support the RS's words. The examples in figure 5.1 demonstrate the differences in wording for a summary comment for a top performer, a middle performer, and a lower-third performer.

The RO's comments, usually briefer than the RS's, will similarly use language reflecting the top, middle, and bottom placement. As the examples in figure 5.1 show, for the promotion endorsement, an RS should carefully employ wording appropriate to the ranking of the individual. Use promotion language along the lines of the following:

Top-third performers:
- "My highest recommendation for promotion"
- "Absolutely must promote"

Middle-third performers:

- · "Highly recommended for promotion"
- · "Promote at the earliest opportunity"
- · "Promotion strongly recommended"

Lower-third performers:

- · "Promote"
- · "Promote along with peers"
- · Don't mention promotion—saying nothing speaks loudly

Not only for this promotion endorsement (or lack thereof) but also in the other details, ensure that the caliber of the comments accords with the intended ranking and the numerical scores.

NAVY EVALUATION AND FITNESS REPORT COMMENTS IN CONTEXT

Over the years, the Navy has upgraded its technology to streamline the administrative process for preparation, review, and approval of performance appraisals, making reviews by selection boards easier as well. Figure 4.3 is a NAVADMIN from 2021 that summarizes some changes to the administration of performance management using eNavFit. Readers should be familiar with the current edition of BUPERS 1610 for following the administrative guidance. On the MyNavy portal, you can review and download the user guide for eNavFit (currently at https://www .mynavyhr.navy.mil/Career-Management/Performance-Evaluation/).

Independent of admin guidance and technical tools, the principles of writing an effective assessment of someone's work performance have not changed. The forms and grading categories for enlisted evals (E-1 through E-6), chief evaluations (E-7 through E-9), and officer fitreps are not identical, yet the writing principles for documenting strong or weak performance clearly for the promotion board apply equally. Like the Marine Corps, the Navy system mitigates grade inflation by showing how the Sailor's average score compares with the RS's historic average. The written comments in an eval or fitrep should be consistent with the numerical scores, promotion recommendation, and milestone recommendations.

Below are some best practices and examples for the written portion of Navy evals and fitreps. What information about the individual's performance should you include? Capturing six months or a year of hard work in the eighteen lines allotted and in meaningful language is a challenge.

Most of the numerical scores on an eval or fitrep do little to get across a sense of the person's real accomplishments or potential. Thus, written comments are essential. By painting a word picture of an individual's strengths based on past accomplishments, a writer can help readers distinguish one person from another in promotion potential. Comments can specify and exemplify important personal qualities and help substantiate and justify the promotion recommendation. They can greatly aid a selection board in distinguishing between dozens or even hundreds of highly praised individuals from many different commands. A person whose write-up does not include convincing, specific substantiation will often suffer in comparison to someone whose write-up does.

EFFECTIVE EVALUATION AND FITNESS REPORT COMMENTS

The primary central content of the written comments should include the following: detailed actions showing evidence of performance traits, the usefulness of that performance to the unit or command, and identification of any special difficulties overcome.

In addition, the comments block should summarize the RS's strongest message to the board about the individual, typically elaborating on the person's ranking and recommending the Sailor for the next major career milestone.

STRUCTURE OF THE COMMENT BLOCK
OF EFFECTIVE EVALUATIONS

"In general, the first line, the opening bullet, and the closing summary

are the crucial parts in evaluation writing. Think of how we read messages.

I read the subject line, the first line, and the last line (the action—

for what the boss wants me to do). In general, with most documents,

we read the beginning and the end, the first line, and the last line.

We don't drop the habit when we drop off the brow."

—CAPTAIN, NAVY SUPPLY SHIP

You have only eighteen lines to elaborate on the numbers and information in other parts of the eval or fitrep. Remember that selection board members often do not have time to read each eval thoughtfully; they must skim. The order of the information can help emphasize the most important details. Common practice is a general three-part structure: an opening summary, a few detailed accomplishments that exemplify key traits, and a closing "bumper sticker" phrase or sentence that expresses the primary takeaway for the board. Consider the following practices as you arrange the statements you have decided to include. Below are several general principles for writing meaningful comments that will help the board decide about promotions.

1. Strong Openings and Closings: Get to the Point Quickly (BLUF)

A short, sharp opening sentence or phrase (not a long generalized paragraph) can set the tone and catch the eye of a reader. It's the place for an important numerical comparison, a vital comparative statement, or an incisive summary judgment. In Blocks 42–43 on officer fitreps and chief evals (Blocks 45–46 for E-1 through E-6 evals), the RS marks an individual promotion breakout—"Promote," "Must Promote," (MP) or "Early Promote" (EP)—and the number of others in the individual's grouping in each of those categories. For a top performer who has only a few others in her category, this block may simply convey that the person is the strongest of three candidates. But in the comments, you can include what is called a "soft breakout," which is a comparison of the person with others of the same rank regardless of specialty. For example, a JO may be the only "Early Promote" out of three other O-3 supply officers (or he may be one of one), but the soft breakout comment in Block 41 can tell the board that he ranks second out of twenty lieutenants in the entire command.

The placement and relative density of information play an important role in making the most important information stand out. By placing information in familiar, expected locations and being careful not to overload the form, one will ensure that readers will see the key comments.

Drafters sometimes wonder about the usefulness of these opening and closing "summary" statements: Won't the details speak by themselves? Keep in mind that selection board members are confronted on report after report with glowing accomplishments, typically filling the form. Over a couple of weeks, they must make judgments on each one of literally hundreds of evals. For maximum effect, the writer should help them assess the meaning of what they see. Some impressive accomplishments carry obvious promotional implications. Others don't—especially to a board member tired of meaningless, unsupported superlatives.

Moreover, Navy board members are long accustomed to looking to the summary of an eval or fitrep to find the conclusive judgment of the commander. The beginning of a report and the end are the crucial locations—anyone skimming or rereading will be sure to look here.

Sample Opening Summaries

The openings of reports listed below (each of which would be followed by specifics in the form of bullets or short paragraphs) all get to the point quickly and are crafted to be noticed.

Exceptionally creative and innovative chief—a real problem solver. My best division chief—already performing at E-8 level.

A top-notch S-3 sensor operator.

A very junior chief who has propelled herself toward the top of her peer group through sheer hustle and enthusiasm.

Newly arrived! Off to a successful start on his first sea tour.

An O-5 unmistakably displaying the early signs of flag potential.

These openings include soft breakouts and other overt numerical comparisons:

My #3 performer among the 6 E-5 RMs at this command.

From a recruit training command:
A SUPERIOR COMPANY COMMANDER who easily ranks in the top 10 percent of 225 handpicked, competitive E-6s.

A superior petty officer and yeoman: #6 of 22 E-6s on STOCKDALE.

#4 of 143 chiefs on THEODORE ROOSEVELT.

These openings denote where an individual fits within a promotion category:

LCDR Smithman, though very junior, ranks #1 of 5 "Promotables."

A rapidly improving leader and petty officer, #2 of 4 "Must Promotes."

Ranked #2 of 6 "Early Promotes"—truly an outstanding chief.

Memorable Closing—Summary Recommendations

A written recommendation, or "bumper sticker" style conclusion to the comments section, serves to restate and reinforce numbers and marks in other areas of the eval form; it can also add some final information. Here are several examples of such summaries, each of which would ideally end the comments section (in some cases the soft-breakout information may be in an opening summary instead).

Summary restating recommendations already marked on the form:
An OUTSTANDING LEADER and TECHNICIAN. Absolutely a "Must Promote" to senior chief. She has my strongest recommendation for selection to the Senior Enlisted Academy.

Summary noting comparative ranking and specific duty recommendations:

My #1 of 7 Petty Officers First Class! Highly recommended for LPO in a communications facility.

Summary that makes recommendations beyond those already marked:

Possessing tact far beyond his years, LT Green would also excel in foreign liaison or would make a superb admiral's aide. Early promote to LCDR!

For a top performer you really want to stand out. The fitrep really needs to say "Early Promote" and be consistent with the breakout of EPs in Block 42 (Block 45 for junior enlisted evals).

Summary explaining a low promotion recommendation:

OS1 White reported on board after four years of working ashore with the Air Force. He shows great industry, has completed 70 percent of ESWS PQS, and is rapidly getting the underway CIC time needed to get back up to speed.

Summary adding one more reason to support the promotion recommendation:

Chief Brown's inspirational leadership is the main reason AIMD New Orleans has the second-highest RFI rate for jet engines in CNATRA. Promote him now to Senior Chief!

Notice how Chief Brown is directly linked to the command's recent achievement.

Candidates who have earned the CO's trust:

LCDR Uniform knows everything about aircraft maintenance, shipboard procedures, and management—I depend completely on his judgment. MY TOP Early Promote!

Mature and confident, HM1 Whisker is my top E-6. READY NOW FOR KHAKIS! Promote to Chief Petty Officer, and detail to the most challenging assignments.

LCDR Top Gun is my #1 department head. I depend totally on her keen judgment. EARLY PROMOTE to O-5 and select for Aviation Operational Command.

An E-6 who has had difficulties in the past:
Petty Officer Victor has overcome a slow start and is performing all his duties very well; he has strong potential as an LPO and technician.

Closing statement after opening summary that says "#3 lieutenant on NIMITZ":
LT Xray has achieved more in this shipboard tour than most officers complete in two. An obvious EARLY PROMOTE! MY STRONGEST REC-OMMENDATION for department head on a major surface combatant.

For someone ranked first of eighteen squadron lieutenants:
LT Zulu is my best pilot. Future fighter squadron CO! Early promote to LCDR and screen first look for aviation operational department head.

One final reason to include such beginning and closing summaries is that after holding an internal board, the command XO may adjust the summaries to ensure that they not only match the promotion recommendations but also are suitably varied from the top to bottom of the group of evals or fitreps the board has reviewed. The tone of the beginning and of the ending, then, are likely to be specially modulated or nuanced.

2. Central Accomplishments: Get Specific

"Everything should be significant. Sometimes the first bullet will speak to leadership. The middle section might speak to something about mission readiness and execution and scope of responsibilities. The last bullet usually ties into something related to professional development or innovation—in evals, perhaps involvement in white hats or the chief's mess—focused on growing."
—NAVY CAPTAIN, FORMER SELECTION BOARD MEMBER

Supreme among the assets of a report writer's skills has to be, now as ever, an able use of specifics. In every service all the literature on performance evaluations cites specific evidence as crucial to a report's success. "Verifiable performance"—illustrated by facts, numbers, and specific achievements—is the key to effective write-ups. Specific details must be foremost. Performance specifics should appear immediately in an eval and make up the greater part of it. If you can quantify it, do so; numbers can speak volumes. "His department scored 97% on the latest maintenance inspection" speaks for itself about job performance.

Naming specific qualities—"leadership," "dedication," "diligence," or "a self-starter," among others—is also important; not everything can be given a numerical score. We all know that the personality and virtues of individuals differ remarkably. Such qualities are extremely important in painting a word picture of the service member.

A major problem in the past was that evals overemphasized personal qualities, often with little or no evidence, thus having minimal, if any, overall influence. This can still happen, but it does not help the service member with the promotion board even if the comments might make the individual feel good when reading them. It's better to focus on a few noteworthy qualities or traits of an individual, citing strong evidence of performance and also assessing the overall positive effect to the command for each one. Support these qualities with evidence; remember, it must support the numerical grades assigned.

Navy eval and fitrep forms underline this principle, calling for "verifiable facts" or "examples of performance and results" and otherwise insisting that the performance "speak for itself." There are several ways in which to do this. The following additional techniques will help make the main content of the remarks section more specific, clear, memorable, and persuasive.

3. Quantify Accomplishments Where Possible

Quantitative measurements are often particularly clear-cut. How many personnel did the officer in question supervise? How many dollars did she manage? Did she save a substantial amount?

This principle also applies for enlisted. How many spaces were repaired and repainted under his direction? How many of his or her junior Sailors advanced on the last exam cycle? Figures, numbers, percentages, dollars, ratios, grades, advancements, and retention—whatever you can quantify—might conceivably be meaningful to a selection board seeking to judge one person against another fairly. Figures are graphic and hard to dispute and sometimes seem to be more objective than descriptive statements. Seek them out and make use of them, within reason and context, with good knowledge of their likely effect.

4. Provide Details about Unquantifiable Accomplishments

Not every achievement has a number attached, but many are equally useful. Did this officer or enlisted complete required professional qualifications? Did this officer qualify as officer of the deck (OOD) and command duty officer (CDO) all on the same deployment? Did a Sailor attain her ESWS qualification in record time or well ahead or her peers? Accomplishments affecting the primary mission are perhaps most significant, and the variations of actual achievements beggar description. Combat experience, participation in major fleet exercises, and hazardous duty illustrating leadership talents obviously are among major subjects for a report, as are important qualifications attained.

Don't overlook mentioning awards, voluntary additional duty, and selections to special assignments or service schools, even if some of these achievements will show up elsewhere in the service record. If you also put them in an eval, they'll be highlighted in both places, and you'll be doubly sure a selection board will notice them. Moreover, the accumulation of several such accomplishments in the space of a year or six months can sometimes significantly impress a reader.

Written or oral commendations from outside the command should also appear. If an admiral or an important organization highly praised the person you're evaluating, mention who did the praising and for what. Also, rather than stating, "This program received high praise from senior officers," name the senior officers or their group. Praise from members of an inspection team could translate to "Fleet Training Group, San Diego, commented . . ." or "'This is the cleanest galley in the fleet,' according to Environmental Preventive Maintenance Unit, Norfolk."

Quotations from flag officers, squadron commanders, or other authorities can be most effective. In the eyes of board members numb with reading volumes of extravagant praise, quotations not only constitute additional evidence of excellence but also add objectivity to the whole report.

5. Back Up Superlatives with Evidence

> *"Give us your opinions, but then give us the data to back it up. Don't just generalize. If a chief prepared the division for a super inspection, what were the inspection results? And if you tell us percentages, tell us percentages of what. To improve retention is fine, but that doesn't mean much if you improved retention from one to two in a department of one hundred."*
> —FORMER SELECTION BOARD MEMBER

Adjectives without supporting details are weak, so support the accolades with facts. There's a world of difference between saying, "This junior officer is an excellent writer," and calling out: "Top Writing Skills. I assigned him to draft all the award nominations, budget justifications, and fitreps we had to do this last quarter on board ship." On the other hand, bullets recording facts without accolades can be equally uninformative. A reader often wants to know not only what but also how the member did. A comment on an officer assigned to BUPERS reading, "She supervised the writing of five precepts for the convening of selection boards," leaves up in the air whether the precepts were any good. But to say of an officer, "He prepared extensive correspondence for SECNAV's signature, all of which was signed without change," suggests the quality was very high.

6. Ensure Continuity in the Report

The comments, soft-breakout rankings, and promotion and future job recommendations should all show an internal consistency and must accord with the numerical scores, recommended career milestones or programs, and formal promotion breakouts. So, if the RS scores a lieutenant as 5.0 across all traits and marks her down as "Early Promote" in Block 42, the write up in Block 41 for this top performer shouldn't just say she is fourth out of ten and call her out as merely "promotable" with no closing summary that recommends department head or future command.

7. Provide Context: Explain Unfamiliar Aspects of the Job

While the eval or fitrep should not be a mere job description, you may have to describe out-of-the-way billets and accomplishments a bit so that board members can justly appreciate job performance.

Ideally, this information will be presented in the "Duties" block on the form (Block 29 on both evals and fitreps). But such details tend to get overlooked, being difficult to read in Block 29 by all those standard listings of every single primary and collateral duty and months spent in each—the board member pressed by time may not look carefully at that block anyway. Beyond that, the eval/fitrep instruction suggests that the "Job Scope Statement" can be continued in the comments block as necessary. So, it's certainly okay to put some job description there. You can also usually work critical facts about the job right into the write-up, often preceding a specific comment with something like, "Responsible for 450 personnel and a $35K consumable OPTAR, she. . . ."

Even though it takes more space, you'll be sure readers will see the description in the comments block. And if your ship or unit took part in some highly unusual activity or faced special difficulty, spell that out in the remarks. The command-employment description (Block 28) may help here. But the relation of command employment to the service member's performance may not be immediately understood; still, many drafters don't have control on what's said in that block, anyway. The bullets shown after the following section helpfully describe the context of a particular job performance.

8. Give Comparisons Whenever Possible

Board members must know the normal range for any provided quantities or percentages to mean something. Without such familiarity or knowledge, the figures themselves will leave them wondering. In many cases a nonspecialist will need some comparison to understand the significance. Where possible, spell out the ordinary standard and then show how this person's work has been specifically above or below that standard.

Occasionally, the comparison will be the ship's schedule: for example, if a chief engineer managed an overhaul so that it was accomplished three weeks ahead of time. Sometimes one compares performance with what people have done in the past, as when a trainer's unit scored 97 percent in Medical Readiness during inspection, which officials said was "the highest ship's score in that area in over two years." For another example, you could say a person increased her unit's Personnel Qualification Standard percentage from 65 percent to 97 percent, or that in a technical area a Sailor "achieved a 98% testbench availability against a Navy average of 87%."

With such comparisons, one gives readers a standard by which to judge the accomplishment.

With these principles in mind, we provide a few examples of eval or fitrep comments. First, consider the poor write-up for a supposedly top performer in figure 5.2. It provides virtually no substantiation for its tiresome adjectives and no specifics connected to the many qualities mentioned. It is also highly repetitive.

At first glance the example in figure 5.2 may sound positive, but not in the context of the eval system. Of course, one can write weakly on purpose; a weak eval

---

FIGURE 5.2. *A Poorly Written Performance Evaluation Aiming to Praise the Sailor*

43. COMMENTS ON PERFORMANCE. * All 5.0 and 1.0 marks must be specifically substantiated in comments.

Petty Officer Smith is an outstanding petty officer and developing technician. He is self-motivated, resourceful, and persistent, projecting a concerned personal involvement in his work duties. An experienced supervisor, he can be relied upon to complete whatever task lies before him or his workcenter and get top-notch results. EM1 Smith in a short time has established himself as intelligent, mature, and experienced, with a strong sense of personal responsibility for the quality of work produced regardless of the job. Manifesting a creative mind and sound judgment, he has demonstrated his capacity for effectively and efficiently directing and controlling the activities of others and for assuring successful results.

He has earned this top evaluation.

> **FIGURE 5.3.** *A Well Written Performance Appraisal*
>
> 43. COMMENTS ON PERFORMANCE. * All 5.0 and 1.0 marks must be specifically substantiated in comments.
>
> Ranks #2 of 31 E-6s in the command.
> **2022 COMMAND SAILOR OF THE YEAR****
>
> STRONG LEADERSHIP AND INITIATIVE! As LPO of his division, consistently maintained the engine room in an "INSURV Ready" status, ensuring all Main Propulsion equipment in top shape. Regularly receives laudatory comments from inspectors about his program readiness.
>
> - As Division Career Counselor carried a first-term retention rate of 85, career 100, and responsible for two first-term BOOSTs, one first-term to sub duty.
>
> - Upon reassignment to "B" Division, completely revamped workcenter EAO 1 by using personnel resources more effectively and initiating a trouble-call system, enabling division to maintain all equipment in commission and "SM 1 Ready."
>
> - Top quality EOOW. Instructs new EOOW's on proper propulsion plant operation, casualty control, and administration.
>
> The epitome of his profession, Petty Officer Brown is a rising leader, has a thirst for knowledge and a desire for challenge. PROMOTE TO CHIEF NOW!

is usually meant to mirror weak performance. In addition to using adjectives that praise him (for instance, "outstanding"), the eval on Petty Officer Smith contains very high numerical marks. But selection board members often discount high promotion recommendations and trait marks if they are unsubstantiated in the comments—as in the case here.

The example in figure 5.3 provides good evidence of sustained, superior performance. Notice how the specific bullets support the opening summary.

Examples of Bullets Citing Specific Accomplishments

The following bullets or phrases demonstrate the various practices explained above, itemizing specific accomplishments, and are followed by (in italics) some comments on their effectiveness.

Noteworthy Bullets

- MMC (SW) Victor's detailed preparations earned high praise from board inspectors during the INSURV and resulted in a superb engineering performance, including full-power turns and emergency crash-back achievements on the first try. *Good specificity.*

- Repeatedly failed to ensure proper endurance of stores. Inability to properly order basic foodstuffs resulted in gross overages of perishable foodstuffs, including enough oranges for six months. *Specifics help substantiate weak performance, too.*

- As Vault Yeoman, instituted major improvements in the handling of classified material, with the result that four other major commands have adopted her specific procedures in their handling of classified material. *Results stretching beyond one's own command are especially impressive.*

- Stands outstanding OOD watches while the submarine is submerged, an exceptional accomplishment as an junior officer. *Notes usually strong performance for her rank.*

- Qualified as OOD Underway onboard DDG 82—first of her peer group to qualify!

- Resourceful, efficient: Recovered and restored the crew's galley and mess deck from severe smoke and fire damage after a mass conflagration on board. Food service management team said it would be "impossible." *The quotation helps make this a telling accomplishment.*

- My only lieutenant of 70 on board who is qualified as an Assistant Command Duty Officer. Routinely stands an impressive watch. *Being the only one out of 70 is tops anywhere.*

Problematic Bullets—Possibly Signaling a Problematic Sailor

The following examples (based on or adapted from actual evals) include weak, ambiguous, or problematic language:

- Supervised numerous unconventional foreign refuelings with NATO forces in Northern Europe. *But how did he do in the supervision?*

- Member of Engineering Casualty Control Team. *This is nothing more than a job description.*

- This Sailor has immediately made a name for himself in the Ready Room. *Is this a positive or negative comment?*

- Reviews all maintenance logs and effects corrective action. *Another unhelpful job description—this would be a good comment, however, if you want to tell the board that this person is completely unremarkable.*

- Her training program was singled out in a successful command inspection as one of the two best training programs audited. *Top two out of how many? And who singled it out?*

- Scored 92.6 at Vieques in October. *The board members may not know Vieques, nor the kind of test or exercise conducted there.*

- Supervised the writing of five precepts for the convening of selection boards. *Yes, but were they any good? Did they yield any measurable results?*

- Thoughtful author. Submitted article to *Approach* on "The LSO's Role in Safety." *A good start, but was it published? If not, can you point to results or in another way capture the value of the expertise suggested here?*

- LT Oscar substantially enhanced office morale by orchestrating and acquiring approval for a plan to obtain new office furniture in order to better utilize available space. *Wordy and unimpressive. Getting new office furniture for a single office isn't a high-level achievement for a lieutenant.*

Set Apart Key Points

Use white space to set apart bullet statements (marked with a hyphen, asterisk, or other mark) to make the most important facts easier to see. Avoid long paragraphs in all eval writing, either breaking them up by the use of bulleted lists or otherwise changing them into short, pithy paragraphs. Bullets also facilitate skimming (and in organizational writing of all kinds, skimming is necessary and expected).

The simple use of some white space to set off key ideas can help ensure that comments are readable, vital information is not buried amid dense blocks of type, and key information is in key places.

Choose the Best Format

As discussed above, a full-block paragraph is possible in the current system, though inadvisable. Also possible are shorter paragraphs or bullets of various sorts. Consider the following approaches to the way to craft and structure the remarks section.

Descriptive Phrases

Another style using bullets begins each with a descriptive phrase. Here, the phrases serve as titles and catch the eye. Figure 5.4 contains an eval on a fire controlman first class written in this format. Note how the use of white space helps set the bullets apart.

The closing comment in this eval is very useful to board members. For one thing, it helps them understand why Golf was assigned as an MAA rather than in a supervisory capacity within his division; for another, it might help explain why this petty officer may not have gotten an "Early Promote" or "Must Promote" endorsement. He's the junior man on board; those senior to him will normally get the nod first.

As you can see, just skimming the "key phrases" by themselves can give a reader a word picture of the individual, and each phrase is supported by a follow-on statement. As for content, note that this eval, though strong, is not exceptional. While Golf apparently is an excellent technician—in another area he sets an example and has "excellent working relations"—but he evidently does not lead.

FIGURE 5.4. *Positive Eval Remarks Using Clear Descriptive Phrases and White Space*

43. COMMENTS ON PERFORMANCE
An excellent petty officer and our #1 MK 56 technician.

* Accomplished troubleshooter. Corrected many casualties during gunnery exercises. Fixed MK4 TDS system that had not worked well in years. (#37)

* An outstanding fire controlman—brought about dramatic turnaround in material condition of gun plot and MK 56 fire-control system. (#34)

Excellent administrator—designed and implemented the annual budget for work center while temporarily assigned as Mess Decks MAA.

Versatile—a great duty MAA; an excellent (temporary) divisional chief; a diligent watchstander: performed exceptionally on gunnery exercise.

Good at attention to detail, greatly responsible for division's 100 percent in 3-M inspection.

PO1 Golf has an excellent working relation with his subordinates. Personally of great integrity—he sets the example for his division.

FIGURE 5.5. *Officer Fitrep with Clearly Written Descriptive Bullets*

41. COMMENTS ON PERFORMANCE
#3 of 15 O-4s across the Command. Unparalleled in technical skill and innovation!

- STRONG LEADER! Takes the misfits and makes them productive. Developed tremendous teamwork between his military and civilian personnel. Personally responsible for energizing physical fitness program with varied yet vigorous training. Extraordinary foresight: developed long-range plans for quality-of-life improvements that yielded FY22 $ where previous attempts netted nothing.

- KNOWLEDGABLE ENGINEER: Formulated and executed award of $3 million window-replacement project. His superbly articulated justification/briefings earned end-year and mid-year $ to complete the project. Junior member of an O-5-level working group: A star in senior role! Managed start-up of design on $30 million MILCON project for a new wargaming center.

- SKILLED MANAGER OF RESOURCES: Executed over 200 transportation and furniture-move requests and 500 maintenance and minor-work requests. Supervised end-stage design and construction start-up of new $200K gym and fitness facility.

LCDR Sierra is a superior CEC officer—already performing at the O-5 level! My top EP.
Screen early for COMMAND!

Action Bullets

Of course, the bullet style is common to naval documents. One standard use is to follow a format common in resumes in which a strong verb begins each of many individual bulleted accomplishments. This method is familiar to naval writers who have often used it on evals, fitreps, and award justifications.

Figure 5.5 is an adapted fitrep for a CEC lieutenant commander written in a similar way.

Once again, the impressive phrases are immediately exemplified—and given credibility—by a wide variety of remarkable statistics. The performance here does tend to speak for itself, which is the ideal of the eval system. Nevertheless, the omission of a summary statement remains problematic. A couple of the less important bullets toward the end might be deleted or combined to afford space for a conclusion.

Fragments

This approach uses one or more paragraphs with a series of several fragments or short sentences linked within each paragraph, one following the other and occasionally interspersing brief adjectives. Navy reports of this kind typically use two or three short paragraphs, grouping like subjects with like. Figure 5.6 shows comments adapted from a fitrep on an officer attached to an NROTC unit, written in this fragment style.

Finally, figure 5.7 provides an excellent example of an eval on a yeoman second class. Despite having only one 5.0 mark, this YN was one of only six individuals out of thirty rated "Early Promote." In addition to the standard eval write-up in Block 43, enlisted evals also include important information in Block 44. This space is meant to include such things as awards received (not just those recommended): honors received, courses or qualifications completed (but not courses in progress),

---

FIGURE 5.6. *Officer Fitrep Effectively Using Fragmented Bullet Style*

41. COMMENTS ON PERFORMANCE

My top-rated lieutenant. An outstanding leader in a large and successful NROTC unit.

*Outstanding MENTOR and role model for future naval officers. Exemplifies Navy Core Values. Honor and integrity unquestionable. Extraordinary commitment to excellence. Moral and physical courage are admired and respected.

*SKILLED MANAGER: Summer Cruise Coordinator for the nation's fifth-largest NROTC unit. Aggressively coordinated and executed summer training for 80 midshipmen dispersed throughout the world. Top-notch Nuclear Field advisor. Designed and supervised extensive, demanding Naval Reactors interview preparations, resulting in selection of seven midshipmen for the naval Nuclear Power Program.

*My finest instructor. A recognized expert in his field. Frequently utilized for difficult, challenging assignments. Extensive understanding of NROTC program administration and operational functions. Maintains a 3.8 GPA in a Master's curriculum in Mechanical Engineering.

EARLY PROMOTE! Screen for Department Head and Command ASAP.

FIGURE 5.7. *Strong Eval Write-Up on a Yeoman, including Qualifications and Achievements*

43. COMMENTS ON PERFORMANCE

Top performing Yeoman, #2 of 6 YN2s on board

On her own initiative, downloaded and installed new evaluation program on 25 computers throughout the command. Assumes responsibility. Enterprising. Can reliably fill any position in the Admin Department and the Chief of Staff office. Self-taught ADP expert—earned apprenticeship based on 2,000 documented hours of data processing.

Monitored awards program/administered LOM/MSM Board—processed 1,000+ awards with less than 60-day turnaround and streamlined board procedures from 60 to 7 days. Self-reliant. Recognized by field and staff commands as the ultimate awards expert.

Education achievements: MS Word Certified, Watchdog Security

PROMOTE EARLY TO FIRST CLASS PO! Ready for LPO of a large division.

---

44. QUALIFICATIONS/ACHIEVEMENTS—Education, awards, community involvement, etc. during this period.

Awarded Navy and Marine Corps Achievement Medal (2nd award). Achieved Dept of Labor, Data Processing Technician Apprenticeship.

---

college credits received, and special certificates awarded. Officer fitreps do not have a separate block for these.

Although the YN report doesn't fill up Block 43, it doesn't need to, because one is convinced by the impressive achievements listed with respect to the striking one-sentence statements "Enterprising" and "Self-reliant" and the extremely strong assessment, "Can reliably fill any position in the Admin Department and the Chief of Staff office." The latter statement indicates this Sailor could immediately take on first class and chief's jobs—probably what the writer intended. And the comment that this YN is "recognized by field and staff commands as the ultimate awards expert" speaks to the great degree of confidence placed by many people in her work and in effect strongly summarizes the eval.

STYLE AND WORD CHOICE IN EFFECTIVE EVALUATIONS AND FITNESS REPORTS

To close this chapter, we offer some tips on word choice to—once again—communicate meaningfully and clearly with the promotion board:

Master (and Minimize) the Use of Adjectives and Other Qualifiers

As long as supporting details back up overall judgments, carefully chosen adjectives and adverbs—terms known by everyone, not just by a tiny group of specialists—can greatly strengthen the overall judgment. Most people are aware of the critical role of adjectives and adverbs in recommendations for promotion; the difference between "he is recommended for promotion" and "he is most strongly recommended for early promotion" is often decisive. But don't overuse these modifiers, and don't use subjective ones that cannot be supported.

Modifiers can help diminish as well as build up. Consider the case of a Sailor who was made personnel officer so that "his daily efforts could be supervised and evaluated." After six months this officer received another eval, one that contained the following faint words of praise:

> In this assignment he has been able to handle most routine office work satisfactorily.

"Most" = not all; "routine" = nothing out of the ordinary; and "office work" = neither management nor leadership nor even physical labor, apparently. As for "satisfactorily," that word itself in an eval will often be interpreted as a warning signal. Clearly, when carefully chosen, adjectives and adverbs can have great effect, for better or for worse.

Avoid Jargon and Unknown Abbreviations

Board members come from all communities. For chiefs' boards, some members are officers, but most are E-8s and E-9s from a great variety of rates and every major community (aviation, surface, and submarine). The one or two members who brief a particular record will usually have good general knowledge of the area of a candidate's expertise but are not necessarily of the same designator or rate.

Write in a way that a generally well-rounded and knowledgeable board member can understand. Don't use jargon; avoid abbreviations that are not known Navy-wide. Don't depend on the reader's knowing all the ordinary measures of success on your platform or in your workplace.

For example, don't write bullets like the following from some enlisted evals, which were obviously written by a specialist and for a specialist—not necessarily for members of a selection board:

- Helped replace the AN/BQN-2 and AN/BRN-3 during ERP.
- Assisted in rewriting of NASNIINST 8025.1 OTTO Fuel instruction and AUW Organization Manual.
- Provided primary RSC Operator Support and SPY-1D maintenance during DDG 72 DT/OT-1/D1 13–19 March 2013.

- Provided primary RSC Operator and SPY-1D Maintenance Support for DDG 73 independent steaming exercise (ISE #1) 12 January–September 2019, (ISE #2) 04 February 2020, and (ISE #3) 18 February–October 2020.

These longwinded titles and figures must be interpreted to someone not knowledgeable in the specialty. The specific dates in the last bullets hold no interest for a selection board member and further clutter the narrative. In addition, note that most of these bullets are once again mainly job descriptions: they outline what was done but not how well.

Altogether, this is the kind of documentation that might belong in a technical report—not in an enlisted eval. Remember to translate technical language into ordinary English, remembering the needs of your readers. Review the guidance on plain language in chapter 1.

Three Final Comments

As with other writing, let the drafts of your evals and fitreps sit a while, then come back to them. You'll be surprised by how different they seem—and by how much better you can make them.

Proofread the comments. These drafts are among the most important documents you will write, and putting mistakes in someone's official file can be both harmful and embarrassing to your people—it conveys that you don't care. Neither you nor they need that problem.

Submit reports on time. Timeliness is a perennial problem with fitreps and evals.

6

Awards and Commendations

"I learned that when I was struggling with the writing of the award,

usually the award was not really merited."

—MASTER CHIEF

EXECUTIVE SUMMARY

People receive personal or unit decorations (usually called simply "awards") and other forms of recognition through the written word. To recognize a Sailor or Marine for heroic action or outstanding work, someone must write down a clear description of what the individual or unit accomplished. Most awards require a summary of action to persuade the approval authority to grant the award. In addition, a citation summarizes the actions more briefly in everyday language so the awardee, family, and friends can appreciate the reason for the recognition. In addition, written letters without a higher decoration awarded can encourage a Sailor or Marine and let others know of their exceptional work. Navy Department civilians can also be recognized with awards. In all cases the most meaningful written documents will provide a combination of general assertions supported by specific details consistent with and appropriate to the level of the award or other recognition.

PERSONAL AND UNIT DECORATIONS

Awards and letters of commendation help build morale on a ship or in a unit. They can also hurt morale if the paperwork snarls and people who should get commendations don't get them, or if the paperwork becomes too burdensome on those who have to write up the awards. Awards are similar to performance evaluations in that they document professional achievement. But their purposes and emphases differ greatly: The performance evaluation focuses on potential for the future, whereas the award nomination recognizes exceptional past service.

The primary Navy document for guidance in writing awards is the *Navy and Marine Corps Awards Manual* (SECNAVINST 1650.1 series). In addition, many commands supplement this guidance with their own specific instructions, providing further details and command-specific administrative procedures. You should find out if your command has its own manual or handbook.

The term "awards" actually includes many forms of recognition. The Navy and Marine Corps (and the other armed services) distinguish between personal (or individual) awards, which recognize the achievement of the service member, and unit awards, service medals, and other forms of recognition, many of which do not involve a formal nomination. This chapter focuses primarily on personal decorations. Still, the guidance below about personal awards applies equally to unit awards. The principles of writing an effective award-nomination package are the same.

Award nominations include two written components: The summary of action (SOA), sometimes called the "justification," and the citation. The audience and purpose for these are distinctly different, as discussed below. For some Navy Achievement Medals and Navy Commendation Medals, the proposed citation also serves as the SOA, and so a separate justification is not required.

For any award write-up, remember to keep the SOA and citation unclassified. Any classified information must be handled outside these elements and in accordance with applicable security measures. The new application system warns users of this.

JUSTIFICATION: THE SUMMARY OF ACTION

> *"The citation is comparatively easy. The hard part is the justification—*
> *selling to the reviewing authority that the award really is merited."*
> —NAVY LIEUTENANT

Audience: The Awards Board

As with every other type of writing discussed in this book, award nominations must be tailored for the intended audience in order to have the best chance of success.

The SOA will typically first go to a board of leaders (officers or senior enlisted) at one's own command. If this board approves the *nomination*, and if the captain of a ship or squadron or a Marine Corps battalion commander does not have authorization to sign the proposed award personally, the package will be forwarded up the chain of command to whoever does have such authority. An awards board at the command, or possibly higher up the chain, typically reviews the nomination and then forwards it with a *recommendation* for approval or disapproval to the person with approval authority.

Thus, the award must often go through more than one committee of approval—a local board first, then to the commodore or higher. Awards requiring SECNAV or higher approval go to an OPNAV board. Obviously, few members of the higher-level boards will know the nominee or the individual's work, so you should ensure the content of the SOA is clear for those outside your command.

Unlike the audience and purpose of an evaluation or fitness report, an awards board will customarily read the whole justification document, not just skim it. They will not, like selection boards for promotion, have to read from five to thirty evaluation comments on every single individual (a quantity that practically guarantees skimming). Awards board members will have time to read the whole SOA. This difference is profound. Looking for patterns under severe time pressure, selection board briefers focus on numerical ratings and the opening bullets and closing comments in evaluations before them and then, if they need to, will scan the accomplishments. In contrast, an awards board will normally turn immediately to the accomplishments to see whether this person's actions or services merit the recommended award.

The *context* for award nominations is also distinctly different from that of evaluations: a person nominated for a personal decoration is not competing against others for a limited number of awards. You must simply show that the individual's actions merit the level of award you are recommending. Rather than having to decide whether, for instance, a particular petty officer first class should take on a chief's responsibilities or an O-5 should receive promotion to O-6, the awards board will be deciding whether past accomplishments merit public recognition.

There is a general misconception that a person must be at a certain rank to be granted a certain award. This is not part of the official policy. But in practice, the scope of action and result required for the different levels of decorations tend to correlate with greater responsibilities, longer time in service, and higher rank.

Also, even though nominees are not competing against each other, an awards board always has a standard to uphold. Giving decorations out too freely can devalue past and future awards. Therefore, the board will consider whether any award is merited, and if so, at what level. If the level of award nominated seems too high, the board may lower it from what you've recommended—very few nominations are bumped up.

Therefore, if you are nominating someone for an award that is higher than typically granted for a person with her or his rank (for instance, nominating an E-7 or an O-3 for a Meritorious Service Medal, which is typically given for higher ranks), you must ensure that the accomplishments and results in the SOA are especially clear and persuasive for that level. In addition to describing the accomplishments very clearly and quantifying or objectifying them in some way, your write-up should repeatedly imply comparisons with the norm and show how this particular individual's service has been consistently above it.

"Frankly, the words often get in the way of actual accomplishments. Puffed up language is a sure-fire indicator of a borderline or unjustifiable nomination."

—PRESIDENT, NAVY DEPARTMENT BOARD OF DECORATIONS AND MEDALS

Content: Writing Persuasive Award Bullets

The electronic space on the SOA form allows for the equivalent of two pages for the justification. Whether you are using an offline fillable PDF or the web-based awards submission system, the quality and clarity of the SOA will ultimately determine whether the award is approved. Normally, except for very high awards or decorations for valor (about which the *Awards Manual* has special rules), the write-up should fit on the electronic form with no continuations.

Opening and Closing Statements

Begin the SOA with a tone-setting opening statement that asserts overall why the individual deserves the award. Remember, the board member usually focuses on the details found in the middle portion to discover just what this person did that might be worthy of special recognition.

Body Bullets

The heart of the award nomination is the body of the SOA. Normally, bullet format is appropriate for this write-up, as it is with evaluations. The central principle here is to cite multiple specific examples of the individual's performance (including names, dates, and places), giving the readers a good general feel for this individual's extraordinary accomplishments. As you evaluate and draft the content, ensure the level of award is appropriate for service rendered and that the SOA supports the recommended level of the award.

Follow these bedrock principles for writing a successful SOA:

- If the nominee had unusual duties, you should explain those duties briefly to make clear the nominee's special performance. The higher up the line the award package has to go for approval, the greater the need to explain unfamiliar achievements.

- For specific content, borrow some of the accomplishment bullets from the individual's recent fitrep or eval (if available). But besides differences in the period covered or the exact events discussed, remember that the aims of the documents differ subtly.

- Make explicit comparisons with the standard or norm, showing how the performance has conspicuously surpassed what would normally be expected of the individual's rank/rate and time in service.

- Delineate specific improvements the service member has brought about by comparing the results of the individual's work with what went before.

- Cite praise from commanders and other authorities whenever possible, quoting the actual words used when they shed light on the individual's performance.

- Highlight personal initiative, voluntary work, and service beyond the call of duty.

- Include the following:
  - Objective, verifiable facts
  - Specific quantities
  - Specific figures, dollars, or percentages when these numbers clarify the special magnitude of the service member's achievement
  - Specific military exercise or equipment names
  - Definitions or descriptions of technical or specialized terms
  - Actions the person took
  - Results of these actions

Avoid the following:

- General or vague statements
- Assessments of the individual's future potential
- Jargon and in-house abbreviations that those outside the command will not understand (less important if the CO is the approving authority)
- Weak verbs (for example, "was," "had")
- A general description of the person's regular duties (they were paid for that—the award is for above-and-beyond performance)
- Subjective descriptions with unsupported adjectives and adverbs

Consider how these examples of body bullets demonstrate the advice given above:

- Personally authored three information one-pagers: "V-22 Osprey Aerodynamics," "Acoustic Tripwire," and "Viking Chainsaw." These articles increased the crew's readiness and were lauded by seniors in the chain of command.

- Repaired a function generator, part of the starboard catapult test station vital to the bench operation, saving the Navy $7,800 in replacement parts and avoiding long testbench downtime.

- Established effective management programs to monitor the processing of over 14,000 demands per month and the reconciliation of over 18,000 outstanding system requisitions, resulting in a savings of $50,000 per year.

- Cited by SURFLANT for "extremely sound judgment and tact."

- The equipment was characterized by inspectors as "the best Basic Point Defense Missile System we've ever seen."

- Made major reductions in the error rate on fitness reports, from the traditional 35–40 percent error rate to 15–20 percent, bringing the command into compliance.

- Reduced the average time per service call from 4.6 hours to 3.5 hours in FY 20, reducing costs by 25 percent.

- Developed three new techniques for torpedo overhaul, resulting in a 60-percent increase in torpedo production, significantly expanding the total number of available weapons for the Atlantic Fleet.

Figure 6.1 offers a strong summary of action justifying a proposed Navy and Marine Corps Commendation Medal. Although based on a nomination drafted before the requirement for a justification at this level was waived, the adapted example still stands as a model of clarity and specificity. This particular medal is meant to be an "end-of-tour" award for a Navy lieutenant after service on board a ship.

A CASE STUDY: WEAK SOA AND A DOWNGRADED AWARD

To further exemplify the process of SOA writing and to show something of its typical context, here is a case study involving an actual award nomination from a several years back. We've changed the name of the officer involved and a few other details, but the basic circumstances are factual. This story illustrates the difference between a weak and a strong SOA and some of the problems of having individuals draft their own award nominations, which is an unfortunately common practice.

FIGURE 6.1. *Strong Summary of Action for a Navy and Marine Corps Commendation Medal*

SUMMARY OF ACTION ICO LT MATT MATTHEWS XXX-XX-XXXX

LT Matthews' performance as Disbursing and Sales Officer during his tour of duty was nothing short of remarkable. Proficient, knowledgeable, and dedicated, LT Matthews instilled a spirit of pride and teamwork in his subordinates, and his untiring efforts led to outstanding results during all of his inspections. There was no job too difficult or tasking too large for him to attack. His accomplishments include:

- Became the first Supply Officer onboard USS DESTROYER (DDG XX) to qualify as both OOD Underway and as a Surface Warfare Officer.
- Passed two major surprise disbursing audits by Defense Accounting Office with outstanding results.
- Ensured the safe execution of 220 flight evolutions during a six-month deployment and several major exercises as Helo Control Officer.
- Was an exceptional DC Trainer, key member of DCTT as Repair Locker Evaluator. His contributions were critical in excellent DC assessments during INSURV.
- Ran a superb Ship's Store. It received a grade of Excellent during the Logistic Management Assessment Inspection and was noted as one of the best Ship's Stores on the waterfront. His efforts greatly increased crew morale.
- Showed outstanding organizational ability. As the point-man in homeport shift, he provided quality presentations and information to over 50 families and crew on the full spectrum of personal and financial requirements and options relating to the Norfolk move.
- Assumed additional duties as Administrative Officer and quickly instituted new procedures to improve effectiveness and efficiency of the Ship's Office.
- Coordinated Combined Federal Campaign resulting in the command exceeding the annual goal by 35%.
- Ranked as my top Division Officer for two years and one of my top underway OODs.

LT Matthews' relentless pursuit of excellence proved invaluable to DESTROYER during his tour of duty. His aggressive, "can do" attitude helped motivate his entire division to a higher level of accomplishment. His performance strongly merits presentation of the Navy and Marine Corps Commendation Medal.

Context

A commander at the Fleet Training Group in Guantanamo Bay told one of the lieutenants to write himself up for a Navy Commendation Medal. Believing his service had indeed merited the award (though feeling some awkwardness about this task), he typed up the cover sheet and composed an SOA and citation (see figure 6.2).

Considering himself worthy and having written up his achievements in such a way, LT North was greatly surprised when he heard later that he would receive a Navy Achievement Medal instead, while another lieutenant, who in his opinion had not done nearly as much, would receive a Navy Commendation Medal. His command's administrative officer later informed North that his write-up simply had not been strong enough. "But I just told them what I did," he replied, "and they know many of these jobs themselves, how important they are. Besides that, the other fellow hyped his collateral duty, and I purposely played down mine—I didn't want it to overshadow the gas turbine engineering work."

Despite LT North's feelings, judging only by the justification, the board may have been right. Many aspects of the lieutenant's accomplishments, factors that explained the quality or magnitude of his performance, failed to come out in the write-up. This deficiency became especially important because, on the day the awards board met, no one from his own department could attend the meeting—sometimes that happens.

LT North sat down much later and considered ways in which he might have clarified the special quality of his work. Below we provide a detailed commentary of how, by explaining his varied achievements, North could have substantially improved the write-up.

FIGURE 6.2. *Weak Summary of Action Leading to a Downgraded Award*

25. SUMMARY OF ACTION

During the period from 30 July 1985 to 31 July 1987 Lieutenant North consistently performed his demanding duties as Gas Turbine Branch Head in an exceptional and outstanding manner. As the senior gas turbine engineer at Fleet Training Group, he was responsible for the Propulsion Systems training on all LANTFLT gas turbine units conducting refresher or shakedown training. Volunteering for the additional responsibilities of command Reserve Training Coordinator, he aggressively pursued a set of initiatives enhancing the utilization of assigned reservists. Lieutenant North's success in this area was evidenced by numerous laudatory remarks and comments by members of the reserve force as well as representatives of COMNAVSURFRESFOR and COMTRALANT. Specific accomplishments during this tour include:

- Supervised the engineering training of 32 LANTFLT gas-turbine ships
- Qualified as Senior Engineering Instructor for DD 963/DDG 993/CG 47-class ships
- Performed duties as Training Liaison Officer on seven LANTFLT ships, receiving numerous letters of appreciation from commanding officers
- Designated Master Training Specialist
- Oversaw the development of Instructor Professional Qualification Standards (PQS) for the Gas Turbine Branch that then became the model for the development of Instructor PQS for all other branches of FLETRAGRU
- Represented Fleet Training Group at the 1986 and 1987 CINCLANTFLT PEB conferences
- Managed the training program for 246 selected reservists performing ACDUTRA at FLETRAGRU
- Researched, organized, and updated mobilization plans and requirements
- Planned, organized, and conducted two successful Reserve Training Conferences for FLETRAGRU reserve detachment commanding officers

Analysis of LT North's Weak SOA

Introduction

The introduction in the write-up does little to set the tone for the whole nomination, includes some unhelpful standard navalese—"pursued a set of initiatives enhancing the utilization of"—and is too long. Readers of justifications tend to be impatient with long paragraphs of introduction, and the subject does not justify the length here, since every item mentioned also appears later in the bullets.

Accomplishments

The set of accomplishments is the key section of the justification comments. LT North's effort is not especially well formatted, and it could use some underlining or boldface type for emphasis. But it does use bullet format, puts the bullets in proper parallel grammatical structure, and makes no errors. Yet the content matters most. Let's go through it line by line and examine how these bullets could be stronger:

- *Supervised the engineering training of 32 LANTFLT gas-turbine ships.* This phrase is simply a "job description," and a very skimpy one at that. While the board members probably have a general idea of what these duties involve, most of them will have no idea of all the specifics and, more importantly, will not appreciate how well North performed the job. In a sense, the lieutenant lost out from the beginning by failing to describe in any detail how well he performed his primary duty.

- *Qualified as Senior Engineering Instructor for DD 963/DDG 993/CG 47-class ships.* This qualification was not a requirement but a voluntary achievement on LT North's part; however, that fact is not obvious. Again, the vital aim of award write-ups is to show unusual, "head and shoulders above the rest" behavior—which this is—so you must take full advantage of opportunities to stand out. The following revision is both truthful and persuasive:

  > At his own initiative, and even though his billet did not require it, he qualified as Senior Engineering Instructor for *Arleigh Burke*–class destroyers. Moreover, he qualified in only four months, rather than the usual eight.

- *Performed duties as Training Liaison Officer on seven LANTFLT ships, receiving numerous letters of appreciation from commanding officers.* We are left in the dark as to what those duties entailed, how difficult they were, and, again, how well he performed them. The write-up does tell us he was commended for them. How much to elaborate on these duties is a judgment call, but at the very least North might have named the ships whose COs wrote letters of appreciation. If he could name three or four specific vessels and even cite key lines from those letters, this phrase wouldn't be a throwaway line (as it almost is now) but a strong justification.

- *Designated Master Training Specialist.* North assumed that the board knew what this qualification means. After all, the same board that approved awards also approved designations, and several months earlier it had approved his. But he forgot that the specific accomplishments that caused them to approve this designation then were not spelled out in writing now. Moreover, by not elaborating on what this designation means, he lost the opportunity to keep the wagon train moving—to show in every accomplishment how superior his performance had been. He could have added significant details, summing up with something like this statement:

  > For these achievements, he was singled out as an expert in training and earned the designation "Master Training Specialist."

- *Oversaw the development of Instructor Professional Qualification Standards (PQS) for the Gas Turbine Branch that then became the model for the development of Instructor PQS for all other branches of FLETRAGRU.* North worked for some time on this particular sentence to see that the value of this achievement came through. And certainly, this is the best sentence in the entire SOA. It indicates that here, at least, North forges ahead and sets the standards for the whole group. Yet the actual circumstances included even more impressive detail, as suggested by the following revision he later made to the last half of the statement:

  > which the group commander personally judged so superior as to make them models for the development of Instructor PQS for all other branches of FLETRAGRU.

- *Represented Fleet Training Group at the 1986 and 1987 CINCLANTFLT PEB conferences.* Once again, don't simply say *what* the person has done, explain *how* the person did it and *why it matters.* LT North realized later that he could have revised this line as follows:

  > Represented the Fleet Training Group superbly at the 1986 and 1987 CINCLANTFLT PEB conferences, despite being the only O-3 among O-4s, O-5s, and O-6s.

- *Served as Officer in Charge of Training Readiness Team sent to the Dominican Republic.* LT North actually left out this bullet from his original SOA at the last moment, perhaps noticing that it was merely a job description. This statement should be included, but it needs more specific detail, such as why he was sent. The revision below would have showcased senior officers' confidence in the lieutenant's soundness of judgment:

  > Sent to the Dominican Republic to evaluate the condition of one of the country's patrol boats and its readiness to go through FLETRAGRU training. His assessment that the boat was "unsafe to steam," and his list of recommended improvements were accepted in full by the Dominican Navy and immediately implemented by FLETRAGRU.

- *Managed the training program for 246 selected reservists performing ACDUTRA at FLETRAGRU.* The lieutenant thought, on reflection, that he could have expanded the reserve-program management alone into many more bullets. As originally written, it simply does not support strong medal recognition. That he managed a training program for reservists, did some research, and "successfully" ran two conferences doesn't say an awful lot. The reported laudatory comments mentioned in the introduction (which the board members have probably forgotten by the time they have read this far) are much too vague to impress. The following revision would have made the bullet and the entire SOA much stronger:

  > Singlehandedly planned, organized, and conducted two highly successful Reserve Training Conferences for 50 reserve unit commanding officers and training officers, O-3 to O-6. In total charge of every aspect of the conferences, LT North earned highly laudatory comments not only from the reservists themselves but also from representatives of COMNAVSURFRESFOR and COMTRALANT.

Conclusion

Although summary comments are not as important in award nominations as they are in fitreps and evals, a strong conclusion can help board members decide what to think about what they have just read. LT North's original SOA lacked a conclusion. A board may even interpret that omission as a subtle signal from the command that they do not wholeheartedly believe in the nominee. Here's a possible concluding statement:

In summary, LT North's service to FLETRAGRU has been highlighted by superb technical knowledge, the utmost of professionalism, and tireless effort. His achievements have always been so far above the norm as to be characteristic of officers far more senior than he. His truly exceptional service would be most fittingly recognized by award of the Navy Commendation Medal.

Of course, not all of the above would fit in a page-long justification statement, but having developed the subject in such a thorough way, one would have more than enough to put together a most impressive justification before deciding which details to remove or shorten, if necessary.

The Revised Summary of Action

The SOA in figure 6.3 revises and upgrades LT North's original submission by adding key information and structuring it so as to catch the eye for greater emphasis.

FIGURE 6.3. *Revised Summary of Action (from Figure 6.2)*

25. SUMMARY OF ACTION

During the period 30 July 1985 to 31 July 1987, LT North consistently distinguished himself in his primary duty as Gas Turbine Branch Head, and also performed exceptionally in a wide variety of collateral and voluntary duties. In this work he:

- Supervised superbly all the engineering training of 32 LANTFLT gas-turbine ships, from each ship's Day 1 Safety Check to its final OPPE Certification.

- Qualified as Senior Engineering Instructor for DD 963/DDG 993/CG 47-class ships, although not required by his billet, qualifying in just 4 months rather than the usual 8.

- Performed duties as Training Liaison Officer on seven LANTFLT ships, receiving LOAs from the COs of the COMTE DE GRASSE (DD 974), STUMP (DD 978), and PREBLE (DDG 46).

- Oversaw the development of Instructor PQS for the Gas Turbine Branch, greatly enhancing readiness. The group commander made this training package the model for all other branches of FLETRAGRU.

- Represented FLETRAGRU terrifically at the 1986 and 1987 CINCLANTFLT PEB conferences.

- Served as OINC of the Training Readiness Team sent to the Dominican Republic to evaluate the readiness of a patrol boat to undergo FLETRAGRU training. His assessment that the boat was unsafe to steam and his list of recommended improvements were accepted in full by the Dominican Navy and immediately implemented by FLETRAGRU.

- Singlehandedly planned, organized, and conducted two highly successful Reserve Training Conferences for 50 reserve unit officers, O-3 to O-6. For this voluntary duty, LT North earned highly laudatory comments not only from the reservists themselves, but from representatives of both COMNAVSURFRESFOR and COMTRALANT.

- Designated Master Training Specialist by board action for excellence in training.

In summary, LT North's service to FLETRAGRU has been highlighted by superb technical knowledge, the utmost of professionalism, and tireless effort. His truly exceptional service would be most fittingly recognized by award of the Navy Commendation Medal.

FIGURE 6.4. *Advice on Writing Award Nominations on Behalf of Your People*

ON WRITING YOUR OWN AWARD—A Navy Commander's Opinion

I have some strong opinions on "writing up your own award." Absolutely, positively, never, never! Every award should be a surprise; the intended recipient should be completely blind to the whole process.

Consider it this way. "Hey, Joe, you're a good guy, how about writing yourself up for the Medal of Honor? I don't really know all of what you've been doing 'cause I'm too busy flapping about other stuff or playing golf [to pay attention to you], but you deserve a medal. Don't worry; I'll make sure you get taken care of."

Later: "Gee, Joe, I guess you're really disappointed about that letter of commendation you got from the admiral after we had discussed getting you a nice medal. Well, maybe they just didn't read between the lines enough. We sent in your write-up pretty much unchanged because we figured you had the best view on what you did. Well, don't worry; it's fitreps and not medals that count, anyway."

Now how does the guy feel? If he had never known what he was being considered for, he wouldn't have been disappointed. Awards are often downgraded, so if you don't tell him he's being put in for a Navy Comm., he won't be upset that he got only a Navy Achievement Medal.

Evaluations are different. The guy will see those in the final form. Here too I never ask my people to write their own, for many of the same reasons outlined above, but I do ask them for inputs, preferably in the form of bullets I can use directly. If the guy wants to write the whole thing, great; but since I sign it, I will edit it—it's never a straight shot from him to signature, even if I just change the "happys" to "glads."

— Navy Commander

For many reasons, having your subordinate write his or her own award nomination is not a good idea. Not only can it encourage self-justification and therefore prove awkward for a fittingly humble service member, but it also suggests that the superior really doesn't know enough (or care enough) to do the recommendation personally. Moreover, it can ultimately prove disappointing to the person you're recommending, as many award nominations are disapproved or downgraded after being submitted. Not least in importance, having a subordinate write up his or her own nomination flies directly in the face of official policy as outlined in the *Awards Manual*, which states not to disclose recommendations for awards either to the recommended individual or to the next of kin.

AWARD CITATIONS
Audience: Non-Navy Family and Friends

"NEVER use an acronym in a formal citation; always use the 'long form.'

Remember the audience (e.g., mom and pop)!"

—NAVY CAPTAIN, SUBMARINER

The citation, beyond almost any other kind of Navy writing, is an eminently public document. Its audience is very different from that of the SOA. The citation is aimed at the nominee, shipmates, children, spouses, other relatives, and friends, many of whom will have little or no naval background. In addition, the citation itself may be framed and hung on the wall or preserved in a special folder and reread for decades—even for generations—to come. Therefore, write it so that it's worthy of such a place.

In addition, make the citation clear for anyone to understand. Use few abbreviations and no acronyms; leave out technical jargon. Make it so clear that it can stand alone when read aloud without requiring rereading or an expert to explain the details. As you draft the citation, imagine a civilian relative or friend listening as you commend the awardee and try to speak to that person.

The citation is brief compared to the SOA. Focus on two or three major accomplishments. Normally, the opening and closing are fixed wording for a given award. You can find those details in the *Awards Manual*. The creativity comes in the middle of the citation, where your words can give readers a sense of the service member's real accomplishment.

Citation DOs and DON'Ts

One of the chief errors that citation writers make is to try to impress with language, using lots of pretentious adjectives and bureaucratese. The write-up should be relatively formal, but formal doesn't mean boring. Do we really want to put family members to sleep by telling them how their loved one "initiated and implemented" something? Or how the Sailor or Marine "developed, coordinated, and executed" something else, even if the person did it in "an exemplary and professional manner"? Doublings and triplings like these are especially common weaknesses of many citation writers.

DON'T use anything like the following:

- Tiresome, long-winded, and pretentious ("boilerplate") phrases such as
  - paragon of success
  - pinnacle of recruiting quality and proficiency highlighted the culmination of
  - in a timely manner
  - coordinated the planning to upgrade
  - was instrumental in the preparation and revision of the driving force in the implementation of
  - to maximize utilization of available assets
  - initiating and implementing an efficient, effective, and accountable organization to gain the necessary visibility and control

- contributed materially to the success
- had a profound influence on the efficiency and effectiveness of
- performed demanding duties in an exemplary and highly professional manner
- (this list is literally endless)
- Jargon, technical terms, or abbreviations that will be unfamiliar to a civilian, such as
  - performed timely recalibration of the MK 113 MOD 9 FCS
  - corrected several difficult problems in the MK 3 MOD 7 Ships Inertial Navigation System (SINS) and the AN/URN-20 TACAN, and completely overhauled and tested all 40 SINS aircraft alignment output terminals
  - performed superbly through a Middle East Deployment, SRA, Operational Propulsion Plant Examinations, NTPI, DNSI, Command Inspection, and INSURV

And don't specify items of interest only to those at your duty station. The rest of the audience won't care that you repaired "NR 1 High-Pressure Air Dehydrator, Numbers 1 and 2 High Pressure Air Compressor, NR 2 Distilling Plant Heater Condenser, NR 2 Sewage Plant Incinerator," and so on.

DO follow these guidelines:
- Use fresh and natural language, make your statement crisp, to the point, and hard-hitting.
- Keep to the facts and let them, as much as possible, impress by themselves rather than depending on the crutch of subjective adjectives and adverbs.
- Be specific enough to get across some sense of what the individual has actually achieved.
- Make all details interesting and intelligible for a civilian audience.
- Write the first and last sentences exactly as specified in the *Awards Manual* for the particular type or level of award.
- Vary sentence style—don't always begin with the subject.
- Write the entire citation in the simple past tense ("improved," not "has improved"; "was responsible," not "has been responsible").
- Use the third person—"he" or "she"—rather than "you" (this rule differs from the style of some letters of commendation/appreciation, discussed below).

- Use the individual's name only in the heading, the second sentence, and the last sentence of the citation.

- Make certain the spelling of the full name is correct, including "Jr.," "III," and so forth, as needed. And use middle initials only—no full middle names.

- Spell out the rank or rate. (See the *Awards Manual* and local instructions for further guidance on "Petty Officer Smith" rather than "Petty Officer Second Class Smith" and others.)

- Capitalize the entirety of any ship's name, and use a ship's designation in parentheses the first time (no hyphen between ship and number): USS MISSOURI (BB 63).

- Ensure all ship and command names and numbers are exactly right. (This might seem obvious, but mistakes of this type are frequent.)

- Read the *Awards Manual*, the basic instruction, and follow the format found there.

- Find and follow the guidance of the local command, the type commander, or any other relevant authority.

- Proofread carefully! You show your own professionalism and the value you place on the awardee by submitting a polished citation.

Citation Examples

The most common Navy and Marine Corps individual awards are the Navy and Marine Corps Commendation and Achievement Medals, commonly abbreviated as NCM and NAM. These do not require a separate SOA, which is combined with the citation in the nomination. Once approved, the citation is submitted as printed on the respective certificate.

This approach not only reduces paperwork but also shortens the length of the actual citation slightly—no more than twelve lines of ordinary type. On the certificate itself, there is room for a citation of only about three sentences between the standard opening and closing language. Shorter may sound easier to write, but the limited space means that it's all the more difficult to make the citation convey something of the actual accomplishments involved.

Figure 6.5 shows a citation gracing a certificate for a NAM that was awarded on board the USS *Los Angeles*. Note that these are typically printed in all caps to skirt difficulties with capitalization. In this example, the standardized language of the opening and closing is set off from the main description by bold italics to make clear what is standardized and what is not.

FIGURE 6.5. *Navy and Marine Corps Achievement Medal Citation*

For professional achievement as material liaison office expeditor, U.S. Naval Mobile Construction Battalion XRAY from December 2020 to July 2021. Petty Officer Yankee superbly maintained an efficient logistical pipeline for five active projects; he personally requisitioned over 1,250 line items valued in excess of $500,000. He also ensured the timely delivery of all required materials and, in the process, earned great respect for the unit by building great rapport with over 25 local businesses. Finally, despite consistently late receipt of funding, he aggressively purchased over $10,000 worth of materials for immediate "startup" projects, thus setting up the command for success during its deployment. *Petty Officer Yankee's professionalism and devotion to duty reflected credit upon himself and were in keeping with the highest traditions of the United States Naval Service.*

Citations for awards other than the NCM or NAM typically run between eighteen and twenty-two lines and must be submitted separately from the certificate. Those for Meritorious Service Medals and more senior awards will typically spell out achievements having exceptionally high value or broad result to major organizations or large geographical areas. The *Awards Manual* outlines (in general terms) the relative level of achievements corresponding to each award.

Figure 6.6 is a well-written citation for an Air Medal for heroic achievement during combat (taken from an unclassified COMFIFTHFLT operation order). Typically presenting details in chronological order, this one exemplifies how such a citation should read so that a layperson can get a good sense of the action involved.

FIGURE 6.6. *Air Medal Citation*

For heroic achievement in aerial flight as a Naval Flight Officer on an F-14A Aircraft assigned to Fighter Squadron FOURTEEN on board USS ENTERPRISE (CVN 65) deployed with Commander, United States FIFTH fleet on 10 October 2001 in support of Operation ENDURING FREEDOM. In the early hours, Lieutenant Doe launched as an Attack Element Lead in the lead section of strikers on highly defended Bar Lock and SPOONREST radar facilities at a vital Taliban airfield. While maneuvering at high speed and high altitude to avoid multiple observed surface-to-air missile launches, guided anti-aircraft artillery and continuous radar tracking, he expertly positioned the F-14A LANTIRN precisely on the target and on time. Under severe opposition, he guided two GBU-12 precision-guided munitions to direct hits, dealing a devastating blow to the Taliban air defenses. Lieutenant Doe also guided two GBU-12 precision-guided munitions from his wingman's aircraft; in all these ways he was a key player in the eventual capture of the airfield, which dealt a severe blow to the Taliban regime. His expert night high-altitude airmanship then ensured a safe high-speed egress from the target area through additional surface-to-air missile envelopes. By his skillful airmanship, steadfast aggressiveness, and exemplary devotion to duty in the face of hazardous flying conditions, Lieutenant Doe reflected great credit upon himself and upheld the highest traditions of the United States Naval Service.

LETTERS AND CERTIFICATES OF COMMENDATION

Navy letters of commendation (LOCs) and Marine Corps certificates of commendation, although not personal decorations, often do carry with them material benefits in addition to simply being valued as expressions of approval. They can add to the multiple for selection through E-6, and enlisted selection boards often take them into account, especially if flag officers sign them.

The format and, to a degree, the style of LOCs vary. Sometimes they are on official parchment complete with seal, command's heading, and other graphics. At other times they are in standard-letter format. When formalized, they usually resemble award citations: you write them in the third person and past tense and use prescribed openings and conclusions.

FORMAL LETTER OF COMMENDATION

When writing these formal documents, follow the same guidance given earlier about award citations: write to a lay audience; avoid hackneyed phrases; describe clearly a few specific, impressive accomplishments; and forgo all jargon and acronyms. Your command may decide to present this letter at meritorious mast or on

---

FIGURE 6.7. *Letter of Commendation Citation*

The Commander, Naval Surface Force, U.S. Atlantic Fleet takes pleasure in commending
MACHINIST'S MATE FIRST CLASS (EOD)
JAMES J. MEAGHER
UNITED STATES NAVY

for service as set forth in the following CITATION:

FOR PROFESSIONAL ACHIEVEMENT IN THE SUPERIOR PERFORMANCE OF HIS DUTIES AS AN EXPLOSIVE ORDNANCE DISPOSAL (EOD) TECHNICIAN WHILE ASSIGNED TO EXPLOSIVE ORDNANCE DISPOSAL MOBILE UNIT SIX, DETACHMENT SIX FROM 8 THROUGH 11 FEBRUARY 2020. WITH VIRTUALLY NO NOTICE, PETTY OFFICER MEAGHER DEPLOYED WITHIN THREE HOURS BY AIR TO COUNTER A POTENTIAL TERRORIST LIMPET MINING AT SEA OF THE U.S. FLAGGED MERCHANT VESSEL "LIBERTY WAVE." FOR THE NEXT 40 HOURS AND WITHOUT REST, PETTY OFFICER MEAGHER ASSISTED IN PREPARING HIS DETACHMENT TO DEPLOY AND CONDUCT AN UNDERWATER SEARCH OF THE VESSEL. DESPITE EXTREMELY HAZARDOUS WEATHER CONDITIONS FOR DIVING AND SEARCH OPERATIONS, HE ENSURED ALL EQUIPMENT WAS ON STATION AND READY WHEN NEEDED. ADDITIONALLY, HE CONDUCTED NUMEROUS PERSONNEL TRANSFERS TO "LIBERTY WAVE" BY SMALL BOAT THROUGHOUT THE OPERATION, AND WAS A MOST INTEGRAL PART OF THE OPERATION'S OVERALL SUCCESS. PETTY OFFICER MEAGHER'S EXCEPTIONAL PROFESSIONALISM AND SELFLESS DEVOTION TO DUTY REFLECTED CREDIT UPON HIMSELF AND THE NAVAL SURFACE FORCE, U.S. ATLANTIC FLEET.

D. J. KATZ
Vice Admiral, U.S. Navy, Commander, Naval Surface Force
U.S. Atlantic Fleet

another formal occasion with friends and family present. Do your best to get across a sense of the actual achievement.

The guidance for writing these letters is the same as for award nominations. The major difference is that you do not have to provide a justification, and the approval process is simpler. Figure 6.7 provides an example of a formal citation from a flag officer. This document (when signed and printed on parchment with seal, force insignia, and other emblems) is appropriate for formal presentation. The write-up presents Petty Officer Meagher's part in the operation clearly and impressively.

Figure 6.8 is the citation for an end-of-tour LOC for an aviation machinist's mate third class (AW). It conveys the achievement effectively (again, a standardized beginning and ending enclose the key descriptions).

FIGURE 6.8. *End-of-Tour Letter of Commendation Citation*

For outstanding performance as Power Plants Technician in Aircraft Early Warning Squadron One One Seven from December 2002 to May 2006. Petty Officer Whiskey consistently performed his demanding duties in an exemplary and highly professional manner. Demonstrating intense initiative and in-depth technical knowledge, he played a key role in the repair of over 2,000 engine and fuel system discrepancies while successfully completing 12 major aircraft inspections. Displaying superb attention to detail, he ensured the meticulous performance of corrosion preventive maintenance, resulting in an overall grade of "Outstanding" during a Commander, Airborne Command Control and Logistics wing post-deployment material condition inspection. His outstanding efforts directly contributed to the squadron earning the Commander, Naval Air Force, U.S. Pacific Fleet Battle "E" Efficiency award for Y2005.

Petty Officer Whiskey's professionalism and devotion to duty reflected credit upon himself and were in keeping with the highest traditions of the United States naval service.

INFORMAL LETTER OF COMMENDATION

To be somewhat more personal and expressive, commanders will often loosen formality just a bit with informal LOCs by using the second person, "you," and by varying from strict past tense.

Compare the fictitious example, adapted from a real letter, in figure 6.9 with the more formal citations in the preceding figures.

The officer who wrote the LOC in figure 6.9 commends the Sailor on his selection as Sailor of the Quarter. The adapted details she cites are subtle ones, harder to "objectify" than many other achievements, yet these can be very important to a customer-service military unit such as a personnel support detachment. Also, notice that, unlike a formal citation, the activities commended here are not all in the past; the use of present perfect (for example, "have led," "have managed," and "have shown") and some present tense emphasize the fact that the Sailor's good works remain ongoing.

FIGURE 6.9. *Informal Letter of Commendation*

From: Officer in Charge, U.S. Navy Personnel Support Detachment Fort Meade

To: PN1 Azuelo Williams

Subj: LETTER OF COMMENDATION

1. With great pleasure I commend you on your selection as U.S. Navy Personnel Support Activity, Fort Meade's Sailor of the Quarter from 1 July to 30 September 2021.

2. The professional and personal traits that have led to your distinction are numerous. Above all, you have managed busy, pressure-filled days with maturity and grace and used quieter times to find innovative ways to improve efficiency. As the person "on the front line," you often provide the first impression customers develop about PSA Fort Meade. Customers consistently comment on their appreciation for your outstanding courtesy and patience, and their respect for your professional advice, information, and quick action on their behalf. Your attention to detail is daily evident in the reports and messages you prepare without discrepancies. You have shown great initiative in designing new ways to improve reporting procedures and keep communication flowing well within the detachment.

3. Your unflagging professionalism has been an inspiration to us all. Well done!

LETTERS OF APPRECIATION

A letter of appreciation (LOA) express thanks and praise for a job well done. It conveys no points, multipliers, or other direct helps to advancement—simply thanks and good will. But an LOA can have another purpose and audience as well: to tell the person's commander or other boss about the good work he or she did. Perhaps a petty officer worked with you on a project but reports to another chief or division officer. A letter to her, routed via her chain of command, will apprise her superiors of her good performance and can raise her standing in their eyes. You can also send LOAs to civilians who give you support via their supervisors.

An LOA rarely needs to be more than one page. A simple three-paragraph structure will help you get right to the point, provide some details, and wrap up efficiently. Open with a clear summary statement (once again, BLUF); you do not need a gradual buildup. For example, in an email to GM1 Adrian Adams, you could open with, "Thank you for providing outstanding technical expertise during our recent weapons refitting." This line immediately focuses the reader's attention and opens the door for supporting details. You could add a small amount of general information to this paragraph, then start a new paragraph to explain the attention to detail, long hours, careful handling of dangerous material, and application of expert knowledge that Adams displayed. The third and final paragraph can simply be a closing sentence or two that reads: "Thank you for your tireless effort above and beyond what was required. Sailors like you are vital to the success of our mission, and I know you will continue to play a key role in the life of the crew."

This three-paragraph formula will keep your thank-you letters short, detailed, and focused on the person being praised. An alternative angle is to address the letter directly to the person's commander and refer to Adams in the third person: "GM1 Adams provided outstanding technical support during our recent weapons refitting."

Figure 6.10 provides an example that demonstrates these principles. The letter tells an aviation structural mechanic equipment airman (AMEAN) that the CO himself took notice of the airman's good work. Written to a junior enlisted man at a very impressionable period of his naval career, it is likely to spur him on to more work of the same and even higher quality. Although the letter is very formal, the commander's sincerity still comes through.

FIGURE 6.10. *Letter of Appreciation*

1650
00/150
24 Nov xx

From: Commanding Officer, Attack Squadron THREE ZERO FOUR
To: AMEAN Jeffery Grant, USNR, xxx-xx-xxxx
Subj: LETTER OF APPRECIATION

1. On 6 October 20xx, while performing a routine turnaround inspection, you discovered a three-inch structural crack in the starboard wheel well of aircraft 407. You promptly reported it to maintenance control, which grounded the aircraft for an in-depth inspection.

2. While the crack was not considered a safety-of-flight structural defect, you exercised sound professional judgment in reporting it. Had this discrepancy been more severe, your sharp eyes and attention to detail might have prevented a mishap.

3. Maintenance men of your caliber are crucial to the safety of naval aviation. Your very commendable action was in keeping with the best of Firebird spirit and the highest standards of maintenance professionals. Thank you for a job well done.

                                                                                       M. R. Khatri

---

As with other letters of thanks (see the section on good-will letters in chapter 2), you should strive to be genuine in an LOA and avoid the appearance of writing just because you feel an obligation. Because a naval LOA usually carries no institutional reward, it will mean little unless the recipient perceives it as conveying sincere thanks.

Civilian Awards

A great many civilians work directly for the Navy and Marine Corps, and others contribute significantly to both services. Here we will briefly discuss awards that may be given to civilians as well as the writing that is involved.

By far the most common civilian awards given out by Navy and Marine Corps officials are those given to Department of the Navy employees. These are called Honorary Awards, and they are governed by the *Civilian Human Resources Manual* and the overarching SECNAVINST 5061.12 series.

As with military decorations, these follow a hierarchy in the level of actions they recognize (and the accompanying prestige) and in the level of awarding authority:

- Navy Distinguished Civilian Service Award: SECNAV
- Navy Superior Civilian Service Award: type commander
- Navy Meritorious Civilian Service Award: CO

Although there is no official correlation between these awards and those given to service members, some experts regard the Navy Superior Civilian Service Award to be the rough counterpart of a Meritorious Service Medal, with the others respectively higher and lower in level.

Non-naval civilians are eligible for three different but parallel awards called Public Service Awards. Like the Civilian Service Awards listed above, these are also titled "Distinguished," "Superior," and "Meritorious," but the award authorities differ.

The procedure for writing up these civilian awards resembles that of other awards discussed in this chapter. Justifications must be submitted along with citations; some recognize specific service, while others honor a long period of exemplary service. The only major differences in the process are that a civilian résumé typically accompanies the civilian award package, and commanders sometimes endorse these award recommendations (via standard letter endorsement) as they forward them up the chain. If written well, such endorsements can help support the written justification. On the other hand, an endorser can also recommend a change in the level or kind of the award.

Like military decorations, civilian nominations involve writing a justification and a citation.

CIVILIAN JUSTIFICATIONS

Justifications resemble the standard Navy award SOA. Follow the detailed guidance provided earlier in this chapter for justifications and citations. Typically written in bullet format, well-crafted justifications will provide many specifics, discuss both quality and quantity, and manifest how this person's service has been consistently above the norm. Within a typical opening and closing, a writer will include bullets like those in the example presented in figure 6.11. The justification was written for a

FIGURE 6.11. *Civilian Award Justification*

From August 20xx through November 20xx, Mr. John H. Smith, the Maintenance Repair Officer for USS YELLOWSTONE, has provided superb support to the ship. . . . Specifically, he

- Crafted the entire work package for a $10 million DPMA for FYxx. The DPMA was instrumental through several SHIPALTS in enhancing the reliability and safety of the engineering plant and Fleet support.

- Coordinated the addition of AFFT stations, Halon and sprinkler system installations in numbers 1, 2, and 3 pump rooms, emergency diesel, flight deck, fireroom, and the engine room. These SHIPALTS modernized the engineering spaces and improved the fire-fighting capabilities in each space.

- Coordinated the replacement of 15 hot water heaters, a critical alteration for the crew's quality of life, done without formal drawings. As a result of Mr. Smith's untiring efforts, the crew enjoys hot water at all hours of the day, including peak usage times. The hot water installation was subsequently praised by the INSURV board as safe and reliable with no discrepancies. A FIRST!

- Expertly drafted and orchestrated a work package for a $1.2 million PRAV in 20xx. Although it was only 4 months in duration, he ensured necessary engineering repairs were made to guarantee a successful Mediterranean deployment. These repairs included the overhaul of the main circulating pump, the trip throttle valves on all 4 SSTGs, the MS-4 limit torque, and the evaporator feed heater condensers. He also saw to the chemical cleaning of both boilers. As a result of this work, YELLOWSTONE is able to maintain both feed and potable water above 90 percent with only one evaporator on line.

- Continuously communicated with the Chief Engineer to provide assistance where and when requested. Mr. Smith continues to stress material improvements in YELLOWSTONE. He truly thinks of himself as a member of the crew . . . and as far as the crew is concerned, he is a crew member.

Mr. Smith has distinguished himself through his diligence, vision, character, and unyielding dedication to the Navy. I have placed my utmost trust in him. My ship's dramatic turnaround and the reputation YELLOWSTONE enjoys today are direct results of his efforts. He is worthy of the Navy Meritorious Public Service Award.

port engineer (a civilian) who managed two extensive maintenance periods on the USS *Yellowstone*.

The endorsement to this SOA (by the *Yellowstone*'s CO) commented on how Smith "fought tough battles" to get maintenance through, despite a climate of fiscal austerity, and how he gained the respect and trust of the crew, becoming "an honorary member of my wardroom" providing superb counsel to the captain about maintenance and repair work.

CIVILIAN CITATIONS

Civilian citations also resemble standard naval award citations. The openings and closings are preset language, and the middle sentences should convey a sense of

The Commander in Chief, U.S. Atlantic Fleet takes pleasure in presenting the
MERITORIOUS PUBLIC SERVICE AWARD to

MR. JOHN H. SMITH

For service as set forth in the following CITATION:

FOR OUTSTANDING SERVICE WHILE SERVING FROM AUGUST 20XX TO NOVEMBER 20XX AS MAINTENANCE REPAIR OFFICER FOR USS YELLOWSTONE (AD 38). MR. SMITH DISTINGUISHED HIMSELF THROUGH HIS DILIGENCE, VISION, AND DEDICATION TO THE SHIP AND TO THE NAVY. HIS EFFORTS IN ORCHESTRATING THE $10 MILLION DOCKING PHASE AVAILABILITY AND THE $1.2 MILLION PLANNED RESTRICTED AVAILABILITY WERE EXCEPTIONAL AND RESULTED IN A SAFER WORK ENVIRONMENT AND A DRAMATIC IMPROVEMENT IN THE QUALITY OF LIFE OF ALL USS YELLOWSTONE SAILORS. HIS EFFORTS ALSO CONTRIBUTED SIGNIFICANTLY TO USS YELLOWSTONE'S MATERIAL READINESS. MR. SMITH'S DISTINCTIVE ACCOMPLISHMENTS AND SUPERIOR PERFORMANCE REFLECTED GREAT CREDIT UPON HIMSELF AND THE UNITED STATES NAVY AND WERE SINCERELY APPRECIATED BY THE MILITARY COMMUNITY.

Signature:
ADMIRAL, U.S. NAVY

the real accomplishment. The writer should target the less-specialized audience of family and friends, avoiding both impenetrable bureaucratese and acronyms. The citation proposed in figure 6.12 (for the same individual whose summary of action is shown above) does pretty well at speaking meaningfully to its wide audience. Again, a Public Service Award citation like this is a citation for a non-Navy employee. Citations for the Civilian Service Awards would be similar; the wording in the opening and closing would differ slightly.

Naval Contests

To end this chapter, let's look for a moment at a related area: special, competitive awards that the naval services have established for both individuals and units. These awards recognize the best performer in a category.

The kind of writing recommended above for standard Navy and Marine Corps decorations—clear writing supported by strong evidence—works here, too. The nomination presented in figure 6.13 is especially well written. The accomplishments are clearly described and most impressive, so impressive that the officer recommended won the Vice Admiral Batchelder Award (given to the highest-performing supply officer) for that year in the "small ship" competitive category. Winning this award was due both to the lieutenant's performance and to that of the outstanding writer who drafted the nomination.

The nomination's author commented that he purposely touched on several different areas of accomplishment to show the nominee's well-rounded qualities. This

---

FIGURE 6.13. *Nomination for a Navy Contest*

LT R—J's performance of duty, leadership, and overall support of this command have been extraordinary. His individual contribution to the supply and operational readiness of this fast attack nuclear submarine has been superlative. Significant specific items highlighting his performance are below:

- Completed an extensive 15-month Integrated Logistics Overhaul (ILO), including significant combat system and nuclear propulsion plant configuration changes. He backloaded the ship's repair parts in only three weeks with a 99+ percent validity.

- Took personal charge of identifying and correcting potential supply support problems as the ship neared the end of overhaul, particularly ensuring COSAL support of several significant ship's systems. In conjunction with this effort, he personally ensured that the entire ship was stowed exactly per plan—an accomplishment unmatched in the Pacific Fleet.

- Prepared, opened, and operated the ship's galley at an extraordinarily high level of efficiency, despite frequent short-fuse demands of shiftwork to support major overhaul events and a severe shortage of mess management specialists.

- Completed Supply Corps Officer Submarine qualifications, an intensive, demanding, and rigorous professional milestone.

- Established a highly effective training program for both MS and SK Divisions and the ship 's RPPOs. Additionally, he provided quality supply input to officer training.

- As the ship's most proficient and professional Diving Officer of the Watch, was assigned to conduct the first dive after overhaul and the first dive to test depth. His performance during these most significant postoverhaul tests was superb.

- Worked diligently to achieve the NAVSEA 08 requirement of 100 percent nuclear (Q) COSAL on board to support the extensive nuclear reactor critical test program. In particular, he achieved this goal without any need to transfer material from any other activities.

LT R— is clearly a most effective and professional Supply Officer. His exceptional work is best measured by the results of the ship 's most recent COMSUBPAC Supply Management Inspection: a perfect score of EIGHT OUTSTANDING GRADES, an achievement unmatched in many years. He clearly merits selection for the Vice Admiral Batchelder Award.

---

approach also helped him avoid technical jargon that would have slowed down the reader. While impressive details fill the entire recommendation, the writer leaves the most stunning accomplishment for the very end.

You can be sure that much of this same material would be used again in an SOA for an NCM or perhaps even a Meritorious Service Medal. While you cannot receive two personal medals for the same actions, winning a contest does not preclude receiving a personal decoration for that work as well.

~ 7 ~

Speaking, Briefings, and Other Presentations

"The ability to speak well, like the ability to write well, will get a junior officer noticed faster than almost anything else."

—NAVY LIEUTENANT, SPEECHWRITER FOR SECNAV

EXECUTIVE SUMMARY

Making oral presentations, often with written or visual supplemental material like PowerPoint slides, is a regular part of the duties of many Navy and Marine Corps personnel, whether enlisted, officer, or civilian. Planning the content and structure of a speech or brief, deciding on the best mode of delivery, and following principles of effective delivery will help make your speeches and briefs more effective in supporting the mission. Here, we first cover preparing your material properly: determining your central idea, choosing the specific elements and structure, and tailoring the words to the situation. We then provide some basic tips on giving the presentation and explain the different tools you can employ to best connect with your audience. Finally, we review three common types of briefs you are likely to find yourself presenting.

THE IMPORTANCE OF ORAL PRESENTATIONS

Wherever you deliver oral remarks, how you deliver a brief or a presentation will often be interpreted—rightly or wrongly—as an indication of what kind of leader or worker you are and of what the people you represent are like. Although we shouldn't live our lives just for appearances, the way we communicate publicly should be consistent with the seriousness and professionalism that characterizes our other duties. In *Naval Leadership: Voices of Experience* (Naval Institute Press, 1987), former Marine Corps Commandant GEN Robert H. Barrow observed:

> If you are testifying before Congress on Capitol Hill, if you speak to the senators or the congressmen effectively . . . , they conclude that the Marine

Corps has good leadership at the top because of the way you come across to them. They think, "This guy must be a good leader because I asked him tough questions and he was forceful and straightforward and forthrightly gave me those answers." On the other hand, if you go out and mumble around, they may never say it, but somehow deep in them they think, "Is that guy a Marine? Is that what Marines are like?"

Speaking with clarity and confidence empowers a leader and can multiply her or his effectiveness in any workplace. This is perhaps truest in the armed forces more than any other setting, given the centrality of leadership and the interdependence between communication and effective command. Whether you are a leading petty officer speaking at quarters, a young officer addressing your division, a commander briefing a combat mission, or an admiral speaking to ROTC midshipmen at graduation from Naval Service Indoctrination, speaking effectively is vital both to mission and morale.

Untold numbers brief their superiors daily. They must prepare informational briefings, decision briefings, reports on deployments, and briefs on the state of a command, to name just a few important occasions. If these speakers don't do these jobs well, the Navy and Marine Corps suffer, either from not having crucial programs approved or simply from not functioning well as armed services. Service members (especially commanders) must also learn how to speak to the public, including local community leaders and the news media, to name just two likely audiences.

Despite this recognized value of speaking skills, we tend to pay far too little attention to the ability to present effectively in front of an audience. College educations often ignore it, for example, and senior enlisted training and officer accession programs may give it only a quick treatment amid the firehose of other topics. This chapter provides some principles to help you begin.

Preparation

"Don't leave the admiral hanging; never give him anything that makes him guess

what you want. Say, 'We recommend this' or 'We want your approval to do that.'

If your brief is only for information, say, 'This is a progress report; we'll have

a recommendation after the final test.' But remember to close the loop."

—COMMANDER, FLAG SECRETARY

ASSESS AND PLAN FOR YOUR AUDIENCE, PURPOSE, AND SITUATION

Know your audience so you can tailor your message appropriately. Why are they listening to you? What questions might they have? What will they care about? What interests them? What is their level of knowledge? Will any topics or remarks offend them? Will they be a friendly audience? You should also find out what officials and senior officers will be there so that you can recognize notable leaders in attendance during your introductory remarks.

Make sure you know whether any Navy or Marine Corps topic will be of interest or if they want you to speak on a specific subject. Also, find out what the group's military interests are. For example, does a local plant produce a weapon system or parts for it? Are local military facilities important to the economy? Discovering such facts might help you tailor your remarks to your audience's interests.

FOCUS AND CONTENT: DETERMINE YOUR PURPOSE AND STATE YOUR MAIN IDEA

Before you can state your main point directly, you need to figure out what your purpose is—what you want to accomplish with the speech or presentation. Doing so is not always a simple task. You should narrow your topic and select the specific content to cover, all with the goal of achieving that purpose for your audience. As you look at your material, scribble notes, and analyze your audience, ask yourself what is that number-one objective you want to accomplish with the brief.

Keep in mind, as well, that your topic is not the same thing as your focal point or "bottom line"—the idea that everything else in your presentation supports. With your purpose in mind, consider what single idea you want your audience to take away when you finish. Do you want to persuade them to agree with you about a position? Are you educating them on a new policy in the office or urging them to take some action? Don't assume your bottom line comes through indirectly or by osmosis, as the saying goes. What one sentence in your speech sums up that core idea?

Structure: Place your Bottom Line Up Front

"It's always important in a brief to put your bottom line up front.

Leave out the fluff."

—NAVY COMMANDING OFFICER

Yes, once again it's the Bottom Line Up Front (BLUF). Just as with so much of the writing we discuss in this book, your speaking and briefing, whether formal or informal, will be much more successful if you articulate your central idea—the focal point discussed in the previous paragraph—from the start. As you structure your presentation, place that statement of your main idea as early as you can. What you want your audience to take away should be clear at the beginning. From there, organize your remarks to elaborate on and support that idea.

Decide on Your Content and Argument Approach

In giving a speech, you may be asked to speak on a specific topic or may have full rein to decide what it is. Even if the general topic is provided or assigned, you still have many decisions to make about your content. You can look into some standard sources for information on general topics. Take the universal military subject of leadership, for example. Whether the occasion is a change of command, an address to a local civic group, or a talk to your own command or division, you need not start from scratch.

As with effective writing, effective speaking requires a healthy combination of general ideas supported with specific evidence or illustrated by details. Without the general concepts, a speech becomes a string of data points or stories without any clear unifying theme or point. Without the supporting details, it can come across as vague fluff that is both unconvincing and unmemorable.

You may decide on your general concepts first (such as four essential qualities in effective leadership or two key steps in ensuring effective system maintenance) and then pull together the evidence and examples to support them. Or you may find it easier to first assemble an array of details and use those to decide on what general concepts to emphasize.

In deciding on specifics to demonstrate and illustrate, you might consult *history* and *tradition*, *literature* and *philosophy*, or *famous sayings*, for instance. All these sources have perennially served military speakers and speechwriters well, as they have politicians. The same can be said for *current events* or *recent military issues*, *war stories*, *anecdotes*, and *memorable expressions* that you have encountered through your service. Consider the following excerpts from naval speeches from decades ago that demonstrate some of these approaches to building or illustrating a speech. In each example, note how the first sentence or two assert a general concept or principle; then the remainder illustrates or expounds on that in the manner mentioned by the heading.

Arguing (or Explaining) from Historical Precedent

[One] goal of American seapower is to be supreme on the sea in order to be supreme on the land. You might recall the historic race for Tunis in World War II, where the Germans moved a quarter of a million troops from France, principally by airlift, into North Africa, but were unable to control the sea in order to supply and re-equip them. The end result was disastrous. On the strategic level, the Germans lost twice: we captured more than 200,000 of their best soldiers when we took Tunis, and we did not have to fight those soldiers when we invaded Normandy two years later. In building our own strategy for the defense of Europe in today's world, we have not forgotten that control of the sea impacts much more than the war at sea.

— SECNAV James H. Webb Jr., speech at the Ninth International Seapower Symposium, Newport, Rhode Island, 28 October 1987, in "Role of American Seapower," *Defense Issues* 2, no. 58, 2.

Citing an Acknowledged Authority

Naval strength is, I believe, essential to America's national security. To keep peace and deter our enemies, we must be able to defeat them if deterrence fails. As Winston Churchill said of his country in another time, "Nothing in the world, nothing you may think of, or anyone may tell you; no arguments, however specious; no appeals, however seductive, must lead you to abandon that naval superiority on which the life of our country depends."

— Secretary of Defense Caspar W. Weinberger, speech before the Dallas Council Navy League of the United States, 7 October 1981, in NL 102 course booklet, U.S. Naval Academy, 1986, 4.

Finding Significance in Recent History

Concerning survivability, just look at this ship [*Nimitz*] which surrounds us with its more than 2,000 watertight compartments, designed and constructed to permit this ship to go in harm's way, to accept battle damage, and to continue to fight. It wasn't designed carelessly or recklessly, nor with bigness for bigness' sake in mind. It was designed with survivability in mind.

Let me remind you that in 1969, the USS *Enterprise* had nine 500-pound bombs explode on its flight deck, the equivalent of being hit by six Soviet guided missiles. Not only did *Enterprise* survive, but within several hours—not days or weeks, but several hours—was capable of conducting flight operations.

It is not unimportant to take another lesson from recent history, remembering that in Vietnam we had over 400 aircraft destroyed and 4,000 additional aircraft damaged—on the ground—on land bases—while not one single aircraft aboard a carrier was destroyed or damaged by enemy action throughout that conflict. Nor should we soon forget that every airfield constructed in Vietnam was lost in its entirety. Even more dramatically, at least one of them has been turned against us, as the Soviet Union today operates with impunity from the field we built there.

So much for vulnerability.

— ADM Thomas B. Hayward, CNO, remarks at the change of command of USS *Nimitz* (CVN 68), 26 February 1982, in "CNO Speaks Out: Why the *Nimitz* Aircraft Carrier," *Surface Warfare*, May 1982, 11.

Linking Tradition to Recent Events

Gustavus Conyngham, the namesake of this fine ship which is affectionately called "Gus Boat," was captured and placed in a mill prison on his first cruise as skipper of a privateer. An individual who did things with pizzazz and a never-say-die attitude, he quickly escaped and was reassigned to a new command. That same spirit is alive today in Gus Boat, in people like MM3 Finan who volunteered to be lowered into a flooded compartment of the STARK with an electric submersible pump which could not be lowered past battle damage by itself. After the line severed on jagged steel, he remained in the compartment, holding down the pump. This heroic action began the dewatering process that stopped the list and probably prevented the ship from sinking.

— CAPT C. K. Kicker, Commander, Destroyer Squadron 2, change of command ceremony for USS *Conyngham* (DDG 17), 14 November 1987.

Illustrating with Specific Details

We need to revise our thinking on ordnance requirements and their associated weapons systems. I turned down some war-winning targets [during the Gulf War] because we lacked a penetrating weapon such as the Air Force's I-2000. Further, the Navy was short of laser-guided bomb kits. In the Red Sea, for example, we started the war with only 112 Mk-82 500-pound-bomb kits, 124 Mk-83 1,000-pound kits, and 258 Mk-84 2,000-pound kits—and that was all we were going to get. The day the war ended, I had 7 Mk-82 kits and 30 Mk-83 kits left. The Navy needs additional laser-guided bomb capability for the foreseeable future and laser spot automatic track designators in all strike aircraft.

— RADM Riley D. Mixson, USN, remarks at symposium in Pensacola, 9 May 1991, in U.S. Naval Institute *Proceedings*, August 1991, 38–39.

PRESENTING

"Military briefers are notorious for abusing visual aids. My rule is: 'If the picture is worth a thousand words, then use it. If not, keep it off the screen.'"
—NAVY LIEUTENANT

SLIDES AND OTHER VISUALS

We've all endured it, even cursed about it: a speech or brief in which the use of presentation slides—usually PowerPoint—became so tedious as to inspire nothing but boredom or confusion.

In a series of routine briefs at OPNAV some years ago, all sorts of mistakes were made. Some briefers had as many as thirty lines (maybe twelve to fifteen bullets) on *each* of their PowerPoint slides. That's far too many to be comprehended readily. Fonts used on some slides were too small to be read from just twenty feet away. Even on "wiring diagrams" (organizational charts that were very important for that particular set of briefs), the print was often tiny.

PowerPoint can make a presentation visually attractive, informative, and unique. Moreover, frequent visuals do enhance a talk, making both for variety and added comprehension. But these visual aids should supplement your talk, not the other way around.

The following are some best practices for preparing and using slides.

Keep Slides Uncluttered

Don't fill slides up with words; three to six bullets are probably enough for any single slide. While you can cram in much more on a slide, it will not always be understood—or even read. Audience members will find it tiresome to be forced to take in large blocks of verbiage "on command," as it were. You won't go wrong if you always remember to have plenty of white space on your slides.

Make Content Central

Avoid the temptation to overuse PowerPoint's many bells and whistles by reminding yourself that content is always central. A brief's slides are commonly forwarded as read-aheads, and afterward officials will make up a missed brief by reading the distributed slides. Therefore, make the materials coherent, informative, and complete. In many cases it will serve these readers better if you give them a more detailed set of slides than the cleaner, simpler slides you use for the presentation. Or you can add as many details into the notes section of the slides for readers to use afterward.

Use Legible, Large Fonts

For most talks, the most readable fonts range around 28–36 point in size, with larger fonts (maybe 40 point) best used for headings. If you are presenting virtually to people who will be watching on their own screens rather than in a large room, you could go down to about 20-point font. Of course, some fonts are more legible than others.

Use Colors That Contrast

The best colors for legibility are black on white. If you prefer color, seek a very sharp yet unalarming contrast, such as gold letters on a blue background. Review the slides personally and with a colleague or two to critique them. But don't just look at them on the computer screen. Find a briefing room with a projector (ideally the one being used for the actual briefing) and view the presentation from the back of the room. Text on a computer screen looks quite different when projected on a screen and read from a distance.

Don't Read Your Slides Word for Word

The audience came to listen to you give a presentation, not to you reading slides. If all they needed to do was read quietly, they could have stayed at their desks. Don't insult their intelligence and bore them by reading for them. Point out the key information or make ancillary points, but let the audience pick up many things for themselves.

Don't Have Too Many Slides

You do not need a slide or a bullet for every single thing you want to say. As a captain in OPNAV commented: "A successful briefer will have five or six slides or viewgraphs, will have thought each one out, and will have written a note to himself about the bottom line he wants the audience to take from it. And if you can't summarize that bottom line, cut out the slide."

Keep Visuals Simple

A graph or chart can enhance comprehension at a glance—which of course is the goal—but some are far too complicated to be understood with only a quick look. In one OPNAV briefing, line graphs were really two graphs superimposed, with one set of values running up the left-hand scale, another set put up on the right, and several lines in the middle. It would have taken ten minutes to puzzle it all out. Another problematic visual was a set of bullets presented right alongside a detailed graph—far too much information for a single slide during an oral presentation.

Proofread, Proofread, Proofread

Pay special attention to spelling, grammar, punctuation, and numbers on slides. Errors stand out even more boldly on a presentation screen and—just as with your writing—can convey the impression of a lack of attention to detail and a lack of general professionalism.

A Final Reminder When Using Slides

Remember, the speaker uses the slides as a visual supplement, but the real presentation is from the speaker. In one OPNAV action-officer briefing, out of twenty presentations (some by admirals), the best was by a chief yeoman. The chief spoke easily from behind a lectern as if she had been born to the manner, with a natural stance and voice. Her PowerPoint slides had about three to six bullets each, all understandable at a glance. The slides were attractive visually, while the bullets themselves were all black on white background with an average 36-point font. Her slides were the most readable, yet they did not oversimplify. And in her presentation, the chief did not depend on the slides to keep the talk moving.

In another example, a one-star admiral at the same venue was giving a talk normally presented by his boss. He used only a single PowerPoint slide, really an outline, for his thirty-minute brief. Yet the admiral made the talk completely his own, illustrating his boss's points with wide and apt reference to Navy procedures and personal experience.

Other Visual Aids

Keep in mind that there are other methods of supplementing your talk with something visual. For example, despite the general preference for slides, a three-star briefing a group at OPNAV instead used a flipchart effectively. His method was to ask what questions audience members had and list those questions on the flipchart. He then oriented his brief around those same queries. This method allowed for spontaneity—and the admiral certainly had the knowledge to respond to all the questions posed. Here are a few examples of additional visual aids you could use in addition to, or instead of, slides. Some of these may seem obvious, but it is easy to overlook them:

- Whiteboard
- Typing on screen (on slides or word processor via projector)
- Flipchart
- Paper handouts (useful if you have a detailed visual to discuss or a lot of detailed information you want people to study afterward)
- An object you can hold up to show or even to pass around the audience

SPEAKING FROM NOTES

Whether you are using slides—but especially if you are not using them—unless you memorize your speech, you will need something to reference as you talk. Speaking from prepared notes or slides is the most common delivery method, called "extemporaneous" speaking. This approach provides some structural security, yet it also allows much more spontaneity than reading verbatim from a text. Notes also enable you to contract or expand the length of a talk much more easily than a prepared manuscript does, if the need arises.

To go about it, type up brief notes to yourself on each area of your talk. Jot down a phrase or clause rather than a full sentence; you can fill in the rest of the sentence as you speak, using a more natural, conversational style. Some people use small notecards or loose paper. Others now like to type notes on an iPad or other device; some even on a phone (which is convenient but may be difficult to refer to easily during your talk). Leave gaps in your notes for filling in details that you know well and can describe easily. You might want to pen in the beginning and ending a bit more fully—these parts of a speech are especially important and deserve more care. Of course, you'll want to make more complete notes for less familiar subjects than for topics you know well.

Don't write out notes for every sentence. Know your material and your outline well enough that you can add explanatory or entertaining details between the lines of your written notes.

Whether you use notecards, sheets of paper, or a mobile device, use large letters for ease of reference as you speak. Typing your notes and printing them in a large font also works well. Make sure to use one side of a page only, and number your loose notes so you can get them back in order quickly if you drop them.

SPEAKING IMPROMPTU, OR "OFF THE CUFF"

Often at a conference or morning briefing, the commander will ask you to "give a rundown" of your program, your latest staff trip, or your plans for the preinspection procedures. The CO might either brief you ten minutes ahead of time on what to cover or ask you on impulse when noticing you across the table. This unprepared, or impromptu, situation has advantages. No one expects you to be polished in such circumstances, and you'll have no time to get jitters. Obviously, you can't prepare much, but you might be able to consider whom you're talking to and what they need, quickly thinking out a brief outline.

Practicing quick-reaction speaking is worthwhile. In a lull at a meeting, recollect one of the tough questions you're currently dealing with and practice putting

> **FIGURE 7.1.** *Improvising with Presentation Context*
>
> ### COMPETING WITH THE ROAR OF THE JETS
>
> Professor Herb Gilliland of the Naval Academy and I once gave a talk at Bolling Air Force Base about a book we had coauthored (a biography of Admiral Dan Gallery, the fellow who had captured a German submarine on the high seas during World War II).
>
> A lecture space in a small building at Bolling had been reserved for the talk on an evening in late fall, but unaccountably, when we arrived there, that room was being used by another group. We were forced to give our thirty-minute talk outside in the dark in fifty-degree weather, competing with loud engine revs from all the landings and takeoffs right across the water at Reagan National Airport.
>
> But we persevered, and the liveliness of the question-and-answer period indicated that the talk had been a success despite all these difficulties.
>
> Clearly, speech conditions are not everything.
>
> Robert Shenk

together a quick mental outline to answer a superior's sudden inquiry. This exercise will help pass the time, and it will pay dividends when someone actually does suddenly put you on the spot.

READING FROM A PREPARED MANUSCRIPT

Of course, reading a manuscript word for word is in some ways the most secure kind of speaking, because you can make sure while composing that you cover everything you want to say and word it just the way you want it. Although this method gives you the most control over what you say and how you say it, be careful—prepared speeches often lose spontaneity. It is easy to slip into an academic, formal writing style that is boring to listen to for a more general audience. We recommend reading from a manuscript only on very formal occasions, and even then, you should work hard to keep the writing in a style that is more conversational to capture the freshness and audience connection that come with something less prepared. In addition, practice reading the text aloud so you get comfortable enough to speak naturally and be less dependent on the script.

VOCAL AND PHYSICAL DELIVERY

Delivery of your speech or brief—successful delivery, that is—goes far beyond making the words in a script or from your notes come out of your mouth. Your entire body and mindset will affect the success of your speech. Consider the following brief (pun intended) advice about elements of delivering your material to the audience in a way that truly draws them into what you have to share.

Audience

We discussed the importance of considering your audience when deciding on your content and the focus of your presentation. The Naval Academy's three-star superintendent demonstrated this principle several years ago when addressing the faculty members, about half of which are civilians. The superintendent decided to forgo flags, the opening ceremony, the salutes, the standing at attention—in short, all the traditional military formalities, including lectern and notes. Instead, he simply stepped through the stage curtains and began speaking. The civilian faculty perceived him as speaking with them rather than down to them. At an institution where rank inevitably plays a great part, this gesture on behalf of the military superintendent to connect better with the civilian faculty was a much noticed and welcome change of pace. It enhanced rather than inhibited communication—and at no cost to the dignity of the admiral or the institution.

Physical Presence

Use your body to engage your audience with what your words mean.

Stance

How you stand (or sit, in a more informal presentation) can affect how both your audience and you yourself feel about the presentation. In most cases it is best to stand straight with your weight evenly on both feet, not leaning to one side or on the lectern. You can convey confidence (whether you actually feel it or not) through your stance. Generally, stand still; avoid pacing back and forth or swaying from side to side. If you are standing still (yet relaxed) for most of your speech, then you can add emphasis by deliberately leaning forward at some point. You can signify a transition of topic by taking a few deliberate (but natural) steps to one side of the lectern.

Gestures

Don't fuss with your hands, but use them to gesture with purpose. Make deliberate, confident movements to emphasize a few key ideas. Make these a bit more broadly if your audience is large. Be on guard against repeating unconscious gestures over and over—these will convey nervousness and distract the audience.

If you don't feel natural gesturing and if doing so would distract you, then leave your hands at your sides or lay them on the lectern (don't use a death grip, however). Don't constantly smooth your hair, grip your opposite arm awkwardly, or make other nervous gestures. Normally, place your notes on the lectern rather than hold them (it's too easy to play with papers in your hands). If you do have to hold your notes, we recommend using small notecards rather than sheets of paper, which can rattle in nervous hands.

Voice

This may seem obvious, but your voice is your instrument for getting your words—and therefore your thoughts—into your audience members' minds. Using your voice well includes the following attributes.

"One of the biggest mistakes is to speed up while reading something to an audience. Perhaps you're reading an award in front of a department. Take the minute to slow down so they'll understand it, rather than just 20 seconds to rattle through."
—LCDR ANDRE LABORDE, UNITED STATES NAVY RESERVE

Pace

Slow down! Many people speak quickly by nature and often speed up when they get nervous. Most of us need to deliberately slow our pace to just the point where it feels a little too slow—that's usually when you are getting it just right. A retired lieutenant commander and former writing and public-speaking instructor at the Naval Academy was known to say regularly, "I have heard countless talks that I found delivered too quickly to really hear everything well. I have almost never heard a talk that I found too slow."

Inflection

Just as you do in conversation, vary your voice pitch and tone up and down at appropriate points to indicate questions, certainty, excitement, or empathy. Think ahead about how to infuse natural emotion into your speech if you want to engage your readers. Act interested and concerned—be enthusiastic and your attitude will come through to your audience.

Enunciation

Carefully pronounce the key sounds at the beginning, middle, and ends of your words. Speaking too quickly can be the enemy of enunciation, so focusing on this element can also force you to slow down as well.

Projection

Speak clearly and loudly enough for the people in the back row to hear you. It's even appropriate to ask at the start whether the audience can hear you there.

Eye Contact

Of course, no discussion of public speaking would be complete without mentioning the importance of eye contact. You don't need to look every single person in the eyes for a long, awkward time, but briefly looking into the eyes of many different listeners as you speak will draw them in and create an atmosphere of a real person talking to real human listeners. Look up from notes, slides, or a manuscript frequently, first looking at one part of the audience, then another. Don't stare, of course, and don't look out the window, over everyone's heads, or down at the floor. Meet the eyes of the people in front of you.

Looking in your audience members' eyes will also help you pick up on how they are receiving the speech. Often you'll find one or two people especially well disposed to you, laughing at your jokes, agreeing with your comments, or simply paying very close attention.

Facial Expression

In addition to the eyes, let the rest of your face play a part as well. There may be natural times in your talk to grin, to look skeptical, to frown, and so on. As a rule, simply smiling or raising your eyebrows as you give your speech will help you draw listeners in and express confidence.

Confidence

You may not feel confident going into a brief, especially if your audience is very senior to you or very large—or both. But working to appear and act confident as you speak will usually make you feel that way as well. Conversely, letting your fears control you will interfere with your connection with the audience, which will then make you feel worse and most likely even less confident. No matter how nervous you feel, often none of your listeners can tell that you're nervous just by looking at you. Even if your voice quavers a bit, people in the audience will usually not notice it as much as you do, or they will just ignore it. Whatever you do, never apologize for being nervous or draw attention to your heightened emotions.

In addition, an apology can make your audience lose faith in you or lose interest in the topic. A former speechwriter for the SECNAV commented: "No speech should ever be self-editorialized. Comments such as 'I really don't know what to say,' 'This is dry stuff so I'll keep it brief,' and 'Thanks for bearing with me' *drastically undermine your effectiveness as a speaker.* Do your best, and let the audience be the judge of the quality of your remarks."

Personal Appearance

Make sure your uniform and overall appearance are top-notch—this is the kind of thing all those uniform inspections were preparing you for. The standard sprucing up is even more important than usual in this context. Plan ahead to have available for you a cleaned and pressed uniform, fresh haircut, shined shoes, straight nametag, and bright rather than faded ribbons—we naturally expect any military representative to look sharp in all these ways. This is not just to impress your audience: knowing you look sharp can help you think sharply and feel prepared and confident. Worrying about a stain on your blouse or how your hair and grooming looks can distract you and drain your confidence.

PRACTICE, PRACTICE, PRACTICE

The way to build confidence and to learn your material is to practice—repeatedly and out loud. Don't just give your speech in your head; you may accidentally skip over the hardest parts and then have difficulty expressing them in the actual talk. Whether you plan to speak from a manuscript or from notes, run through the speech out loud several times in private, perhaps in front of a mirror or camera, even both.

If you can find friends or colleagues to listen to you, practice in front of a live audience. If possible, choose someone who is knowledgeable in speaking and helpfully critical. Often, naval staffs will "murder board" a critical briefing before the toughest local audience they can assemble. The tougher the questions you face now, the better prepared you'll be in front of a target audience of major officials later. Such a critique, in which briefers and staffers try to anticipate and practice answering likely questions, can be an outstanding learning experience in its own right.

Try recording yourself with your phone or webcam and then give yourself a constructive critique. You'll probably feel embarrassed the first time you watch yourself, but this experience can be absolutely invaluable for your development as a speaker.

Naval Briefing Types

The principles and techniques of effective speaking apply to military and naval briefings just as to any other type of speech. But military briefings are distinctive in several ways. Usually, they are relatively brief (indeed!) and to the point. Because the audience is typically a "command audience," comprising mainly insiders—a broad category we often call "staff briefings"—briefers don't usually need to use attention-getting devices. You also seldom need to explain terms and concepts. In addition, many briefings are directive or informative in nature, so you have less need to persuade than you would otherwise.

Still, be prepared to demonstrate your expertise and even defend your position. Disagreement is possible in many more situations than you would expect. Usually, you will be briefing an issue or recommendation because you're the expert on the subject. The credibility so important in writing applies to speaking as well—perhaps more so. Therefore, know your subject backward and forward and prove your expertise when someone challenges you during a brief.

Below we discuss some distinct types of staff briefings and offer advice for preparing them.

INFORMATIVE BRIEFINGS

Many briefings exist simply to keep the CO and the staff informed about command operations. Besides ensuring an exchange of information among staff members, they offer opportunities to announce decisions, issue directives, share information, and give out general guidance. These purposes serve a command's larger goals of unity and coordination.

Many variations of informative briefings exist. One officer may brief the entire staff, or several staff officers might speak in succession. The method and the formality of such a briefing will depend on the size of the staff, the nature of the command, and custom. The XO or chief of staff will usually preside and set the agenda, and the CO will normally conclude the briefing. Both officers may take an active part throughout the presentation.

Informative briefings are often made to a staff about a particular event or mission although they may be more formal and deal with only one rather than many issues. All such briefings deal mainly with facts rather than recommendations. Obviously, details will vary with topic and situation, but here is a recommended simple, standard speech organization:

Introduction

- announce the topic
- orient the listener about why they need to hear the brief
- state the main point (BLUF)
- briefly summarize the structure of the main body of the talk

Body

- present facts in an orderly, objective, clear, and concise way

Conclusion

- reiterate the main point and supporting ideas
- emphasize why the main point matters
- reinforce any action or implications of the main idea

Try to anticipate questions that might arise and treat most of them in the briefing itself, before the audience asks. Bring along any background information that might help you respond to questions but know your subject so well that you can respond directly to any reasonable inquiry.

Of course, some commanders will use staff briefings to help them make decisions; in that case the presentation becomes a decision briefing.

DECISION BRIEFINGS

You will have many occasions to advocate one course of action or another. In some cases you may not personally prefer one decision over another, but you'll often be making your case before a superior who has the responsibility for decision. This presentation might be in informal circumstances, perhaps while standing before your boss's desk, or on a very formal occasion, while giving a decision briefing before an officer of flag rank and several members of that officer's staff.

Whatever the circumstances, keep in mind that several factors other than what you actually say will influence the outcome of your briefing:

Your credibility with the audience. Even if everyone on the staff knows you, they may not all know the particular expertise you bring to the issue. Have you, or has anyone, explained to those you're briefing what your background is or why they should listen to you? An introduction that covers that will go a long way to persuading them to agree with you and to trust what you say.

Your rank. You may have the knowledge and experience, but how does your rank affect the way you are perceived? If you are junior to your audience, you may need to work harder to establish credibility.

Relationships with the audience. Might some in the audience be antagonistic to you, to your office, or to your topic for any reason? If so, anticipate how you might deal with opposition. On the other hand, do you have allies? Will anyone in the audience help you? How can you best draw on such supporters?

Physical context. Will you be speaking in your own offices, on your own turf, or via a virtual platform like MS Teams or Zoom? Will you brief in a large room or a small one? Are you speaking to a large group or at a small, private meeting? Will the audience sit around a conference table or face you in classroom chairs or auditorium seating? Although these factors may not change your primary message or structure, you may need to adjust your tone, physical stance, presentation slides, or other elements based on this context.

Timing. Do you have enough time to say what you plan to say? Will another evolution interrupt? Will people be distracted by where they need to go next if you go over time? Also, remember that the time of day (early morning, just before lunch, midafternoon, and so on) may influence the mood, attentiveness, or even the wakefulness of your audience.

In some settings you can expect a decision on the spot. In fact, the flag or flag's staff may have asked for the decision brief for the very purpose of deciding the issue quickly. In that case you might end your brief by saying, "General, I have completed my presentation, and I am prepared for your decision," or words to that effect. This has been standard procedure on some Marine staffs but is less common with Navy staffs.

Yet as the briefer, you shouldn't try to force an immediate public decision. If the senior renders no decision immediately, leave the issue at the "recommend" level and pursue the decision later through staff channels. Let the senior officer's staff guide you in how to proceed.

Figure 7.2 presents a proven format for a formal decision brief with guidance. It is based on the classic staff-study format as adapted to oral-briefing requirements by the Marine Corps Development and Education Command at Quantico, Virginia.

Again, the format in figure 7.2 is simply one model or template that many have used in the past. Alter it as you wish based on your particular subject matter, the occasion of the brief, and your knowledge of the decision maker.

MISSION BRIEFINGS

Service members deliver mission briefings to Marines, ship drivers, and naval aviators alike as they are about to embark on operational missions either for training or for actual combat. Although such briefings will vary widely in technique, subject, and location, they each have one central aim: to instill the best possible understanding of an impending operation in all participants.

For example, a Marine patrol going out will normally learn of its specific mission through oral orders. Then a mission briefing may provide further specific instructions, such as the route of march, what to look for, identification procedures, and so

FIGURE 7.2. *One Format for a Decision Brief*	
FORMAT	**EXAMPLE**
Greeting. Use military courtesy. Address the decision maker and other key persons in the audience. Identify yourself and your organization, if necessary.	Good afternoon, General J——. I'm Colonel M——, the Staff Operations Officer.
Type of Briefing, Classification, Purpose.	This is an unclassified decision briefing.
Subject and Problem. State very briefly the background and present context of the problem at hand.	As you know, one Marine per year is killed in the minefield at Guantanamo Bay, Cuba, and incidents are occurring more frequently. Lately, this problem has attracted congressional attention.
Basic Recommendation (BLUF). This is perhaps the most important advice of all. Don't leave your listeners in any doubt about your basic recommendation. Tell them early and speak confidently.	To solve this problem, I recommend that we replace the conventional munitions in the minefields at the U.S. Naval Base at Guantanamo Bay with FASCAM munitions.
Detailed Statement of the Problem. If necessary, outline more fully the problem this briefing intends to solve.	Defense of the U.S. Naval Base at Guantanamo Bay has been a significant issue since 1959. Much of the existing barrier relies on antitank and antipersonnel mines. We must emplace and replace these conventional munitions by hand. This process is time consuming and dangerous...
Any Necessary Assumptions. State any assumptions needed to bridge gaps in the data. Make sure they are reasonable and be prepared to support them if challenged.	We can assume that the political situation will remain the same for the foreseeable future and that the mine barrier must continue to remain in place. Thus, an improved process to emplace the mines would seem a long-term need.
Facts Bearing on the Problem. State pertinent facts objectively. Present both sides of the issue, even if recommending just one. Research indicates that a high percentage of audience members will lean toward your argument when you present both sides, but just a few will when you present only your side. Be sure to cite authorities and relevant supporting opinions.	1. FASCAM munitions now available offer some antipersonnel, antitank capabilities with distinct improvements over conventional landmines. . . . 2. FASCAM munitions can be emplaced by artillery and so offer a significant safety improvement. . . . 3. FASCAM munitions allow for a more flexible response because they can be fired in reaction to enemy action. . . . 4, 5, 6, etc.

FIGURE 7.2. *(continued)*	
FORMAT	**EXAMPLE**
Possible Courses of Action. State major feasible options. Explain the advantages and drawbacks of each and any potential dangers involved.	1. The major alternative to using FASCAM munitions is the current method, which is to bury mines below ground level by hand, a tedious and dangerous process. With this method, we must painstakingly record mine locations and replace each mine before its shelf-life expiration date. . . . 2. The major advantage of the present system of mine emplacement is cost. Equivalent mine munitions are considerably less costly than FASCAM rounds, and FASCAM will require augmenting the Security Battalion with a 155-mm howitzer battery. . . . 3, 4, 5, etc.
Analysis. Present your conclusions briefly. Mention any concurrences and nonconcurrences.	1. The admittedly significant increases in cost, personnel, and equipment are worthwhile when weighed against the recurrent loss of life presently incurred in handling conventional mines. The employment of FASCAM will totally eliminate this loss of life. 2. Besides being safer, FASCAM minefields are more effective and more flexible. 3. CMC received a brief during a visit to GITMO and liked the idea of FASCAM
Restated Recommendation. Restate your recommendation, wording it so it requires only approval or disapproval.	We recommend that FASCAM munitions replace the current mines emplaced at Guantanamo Bay.
Opening for Questions. Try to anticipate questions; conduct murder boards if you can. Do your best to have thorough knowledge of the whole issue so you can respond intelligently to questions or arguments.	Are there any questions?

on. It may also afford the Marines a brief explanation of why the patrol is necessary and what it will contribute to the overall mission of the command.

Ships have similar procedures. Key officers and senior enlisted personnel on a destroyer, for instance, might go over to the flagship by helo for a mission briefing on an impending operation, perhaps an operation involving naval-gunfire support, air operations, underway replenishment, or a missile shoot. The briefing would include all manner of specifics, from call signs, communication details, and emergency procedures to formations, tactics, and weapons employment. In addition to providing specific plans and details, the briefing officer typically would touch on how the operation fits into larger operational or training objectives.

STAFF BRIEFINGS

Figure 7.3 is an extended, excellent example of notes for a staff briefing about briefings. This comes from United States Navy Strike Fighter Tactics Instructor Program (formerly Naval Fighter Weapons School, or TOPGUN), which as part of its training program long ago developed a strong program of operational briefings and debriefings. Naval aviators pay great attention to effective delivery of mission briefings, partly from having to give them so often. These excerpts were adapted from a version of the school's guidance. Although originally designed to aid those who must brief Navy and Marine Corps fighter squadrons that are about to fly a tactical exercise, this advice and structure are adaptable to many other situations.

Take note of the advice given, much of which echoes the guidance provided earlier in this chapter, though as you'll see, it's tailored for the particular command and situation of mission briefings at TOPGUN. As such, it is a model of adapting general guidance to specific contexts and audiences.

FIGURE 7.3. *Staff Briefing from Top Gun*

Briefing and Debriefing at TOPGUN

A. **Introduction**. Tactical flight time will continue to be at a premium in the months and years ahead because of funding constraints and aircraft availability. We must take advantage of every opportunity to refine our aviation skills. Your squadron has made a large investment in OPTAR, TAD funding, and maintenance support to provide for your attendance. It expects a return on the investment. One way you can pay off is by giving professional briefs and debriefs. Comprehensive briefs and debriefs are the cornerstones of an effective squadron training program. While you may not use all these pointers on every sortie, they apply to almost any tactical fighter mission.

B. **Preparation**. If the first time the briefer has considered the mission is thirty minutes prior to the flight brief, then the briefer has done a disservice to the squadron by not taking full advantage of a valuable training opportunity. Mission planning varies widely, but certain elements are applicable to all missions.

    1. As flight leader, have a clear idea of what the mission and/or training objectives are. If the flight leader doesn't have it clear, certainly no one else in the flight will.

    2. Start mission planning early—at least the day prior.

    3. Involve other members of the flight. This will be no problem if the mission is an air superiority sweep in the Gulf of Sidra but some arm twisting may be required for night max-conserve 2v2s off the coast of Diego Garcia. Make sure all members of the flight arrive at the brief having familiarized themselves with the SOP, mission objectives, operating area, and any other information required to maximize the performance of the aircraft. Anything less is unprofessional.

FIGURE 7.3. *(continued)*

4. Develop a scenario that will be challenging but within the capabilities of all members of the flight. Don't give a lengthy dissertation on a country's political-military situation, but rather make the situation a detailed framework from which the fighters can make intelligent tactical decisions.

5. Review written material for guidance on tactics, maneuvers, and the threat. Refer to Tactical Manuals, TOPGUN Manuals/Journals, VX-4 Newsletters, and related materials for information. Dust off the contingency plans from the last cruise or deployment for reference.

6. Allow for contingencies such as maintenance problems and weather aborts. Plan an alternate mission.

7. Write it down. Putting the brief on paper—either in outline or paragraph form—will help briefers organize their thoughts and make them more familiar with the material. The result will be a smooth brief that is not repetitive and disjointed. If the brief is large and complex, a practice run-through is worth the effort.

C. **The Brief**. Below are some specific pointers on the brief itself.

1. WHERE. During shore-based operations, it's not too difficult to set aside a briefing/debriefing room—complete with whiteboard, models, and big screen—that allows for a quiet atmosphere without interruptions. Unfortunately, the reality of shipboard life is such that Navy and Marine Corps squadrons must often conduct their briefs amid the confusion of the all-purpose ready room. The squadron duty officer must ensure minimal interference with a flight that is briefing or debriefing. Try posting a sign on the ready room door that alerts everyone that a brief or debrief is in progress.

2. WHO. The flight leader traditionally conducts the brief. However, briefing can provide valuable training to a less-experienced member of the flight. Naturally, any briefer should have been intimately involved in the planning.

3. WHEN. Most squadrons brief 1–1½ hours prior to man-up, depending on the mission and size of the flight. You'll have enough time to cover all the necessary items—if you're prepared. When the squadron is operating in a new locale you may have to lengthen the brief to cover local course rules and operating areas. DO NOT SACRIFICE THE TACTICAL PORTION OF THE BRIEF FOR ADMIN ITEMS.

4. SETTING UP. Here are a few general pointers:

- Be at the squadron early to take care of any last-minute items such as aircraft availability, weather, scheduling changes, etc.

- Set up the briefing room so that everyone has a full view of the briefer, whiteboard, and other briefing aids. Clean the whiteboard of all items not pertaining to the mission. Check that models and colored markers are available. If time constraints prevent whiteboard preparation, have the briefing items photocopied and passed out.

- Start on time! A flight that briefs late will walk late, take off late, etc. It only takes one instance of losing a hop for tardy players to change their ways.

- As briefer, remain standing throughout the brief. This posture makes for better delivery and reinforces the briefer's leadership.

- If you have players who were not involved in the planning, start with a brief overview of the mission.

FIGURE 7.3. *(continued)*

[The TOPGUN instruction continues with details of the "ADMIN Brief" on take-off times, comm plan, weather and divert procedures, mission and training objectives, details of the "TACTICAL Brief" (the heart of the mission), and the set-up of each engagement. Then the instruction continues as below.]

5. BRIEFING TECHNIQUES. All of us have our own briefing styles. Note others' effective techniques and use what works well for you. Here are some suggestions:

- Keep an eye on the clock and pace yourself during the brief. Allow ten minutes at the end for questions, crew coordination, and a pit stop prior to man-up.

- Maintain eye contact with the flight members. It will keep them attentive and provide feedback as to whether your points are getting across.

- Ask questions from time to time to keep everyone involved in the brief. They can be rhetorical or specific. However, be careful not to bilge your wingman by playing NATOPS Trivial Pursuit in front of the CO.

- Recap the mission in general terms as a conclusion to the brief, with a review of the training objectives.

- If you complete the brief with time to spare, cover any tactical contingencies/issues that are relevant. As Navy and Marine Corps officers, we accept the paperwork burden as an unavoidable price to pay to fly high-performance fighters. Don't sacrifice a scheduled opportunity to talk tactics just to read the message board, make a phone call, or push papers between the brief and strapping on the jet.

[A section on "Remembering the Flight" comes here.]

D. **The Debrief**. The most important consideration about the debrief is to have one! We neglect or gloss over many debriefs because of follow-on missions, lack of space availability, crew rest, or a hundred other reasons. Too often, we lose valuable training/learning because we're not interested enough in conducting a meaningful analysis of the mission. If the mission was so mundane or routine that it doesn't merit a debrief, then the flight lead was negligent in identifying training objectives—and then you should note that fact in the debrief.

The debrief should not be simply a chronological regurgitation of the mission. You should analyze how well the mission objectives were accomplished, discuss what went wrong and why and what went well and why. You should also identify lessons learned for subsequent missions.

The debrief begins as soon as the brief is over. Jot down any point worthy of discussion. After the mission, review your notes and recordings to organize your debrief comments. Ask yourself some pertinent questions: Were the mission objectives achieved? How about the training objectives? What mistakes were made? Were the mistakes due to poor planning, briefing, or execution? A few minutes taken to organize your thoughts will dramatically improve the quality, expeditiousness, and professionalism of the debrief.

Follow these additional guidelines:

1. Have the debriefing room set up with whiteboard, colored markers, models, and a big screen. Draw notable geographic features of the operating area on the board, with north oriented to the top, and include the position of the sun. In one corner of the board list all the players next to the color of the arrow that will represent them, and list the training objectives as well.

2. Make sure all players attend the debrief, including the controller and the adversaries.

> **FIGURE 7.3.** *(continued)*
>
> 3. The overall debriefer—generally the flight lead—is responsible for maintaining control of the debrief. Emphasize that everyone will have a chance to talk, but only after the debriefer has first addressed the important points.
>
> 4. Spend the first few minutes of the debrief covering any ADMIN problems (clearance, line procedures, rendezvous, recovery, etc.). Get these matters out of the way quickly to clear the air for the important TACTICAL debrief.
>
> 5. As the debriefer, actively promote an atmosphere that encourages frank discussions without recriminations. See that all participants set aside personal feelings, friendships, and rank as they walk in the door. At TOPGUN we recount engagements in the third person. Instead of, "Here's where I gunned you, Dirt, when you were obviously out of knots and tried a nose-high guns defense," say, "At this point the attacker achieved a valid gunshot when the F/A-18 attempted a nose-high guns defense at a low airspeed." Both examples address the mistake, but the players will accept the lesson more readily in the second case.
>
> 6. Solicit input from the crowd to keep everyone interested in the analysis. Admit mistakes to maintain credibility, and acknowledge good performance to reinforce the positive. Keep the discussion oriented to the mission objectives and relevant points.
>
> 7. Structure the TACTICAL debrief to cover the important points thoroughly. You can address relatively minor points that didn't affect the success of the mission at the end of the debrief, or perhaps not at all.
>
> [The TOPGUN instruction goes on to cover methods of boardwork in drawing fighter engagements and details of the TACTICAL debrief. It then concludes this section.]
>
> 8. After the TACTICAL debrief, summarize the flight with reference to the training objectives identified in the brief. At TOPGUN we use a "Goods and Others" format. This discussion provides the basis for determining training objectives for subsequent flights.
>
> **E. Summary**. The trademark of the TOPGUN graduate is being the best briefer and debriefer in the squadron. In the next five weeks, carefully observe the techniques of all the instructors. Select what works the best for your own personal style, and perfect your briefing and debriefing skills. Lessons learned in peacetime training must equip us for the challenging scenarios we can expect in modern aerial warfare. In the final analysis, the aircrew's flying skills will determine success or failure of the mission, no matter how well equipped they might be with weapons, intelligence information, and policy guidance. These aviation skills are a direct function of how well you've briefed, led, and debriefed your aircrews.

Closing Encouragement

In the naval service your communication skills will distinguish you from others more than just about anything else. Enlisted and officers in positions of responsibility and expertise can expect to speak to groups regularly as part of the job. Speeches and briefs merit sound preparation and skillful execution just like everything else we do. You can become skilled at public speaking by knowing its principles and by practicing. Work at it until you can do it well and quit worrying about whether you're up to it or not.

~ 8 ~

Technical Writing

"One of the greatest writing challenges I see among professionals is explaining and translating material they understand well for readers who don't."

—FORMER WRITING CENTER DIRECTOR, U.S. NAVAL ACADEMY

EXECUTIVE SUMMARY

Technical writing adapts specialized content to specific readers and purposes. Although many types of writing fit this description, here we cover three specific documents (or elements of a document) Navy and Marine Corps writers may find themselves preparing: the technical report, executive summary, and abstract. Understanding the purpose of each of these and the perspective of the audience will help you make them more engaging and effective. Principles of plain language apply to these document types; the technical writer should aim to make the style as clear and appropriate for the audience as possible.

WHAT IS TECHNICAL WRITING?

"Technical reports are meant to be skimmed! Hence the frequent occurrence of executive summaries, abstracts, section summaries, appendices, frequent headings, [and] even summaries of summaries."

—FORMER PROFESSIONAL WRITING INSTRUCTOR,

U.S. NAVAL ACADEMY

Technical writing can refer to a broad range of communication and document types. College textbooks on technical writing cover everything from emails to equipment specifications to recommendation reports. Whenever you are composing specialized information for a specific purpose and audience, tailoring what you say and how you say it for the situation, you are doing technical writing. In fact, everything covered in this book could be considered technical writing insofar as it is about writing in specialized naval (or Marine) contexts for specific readers in real-world situations. In other words, technical writing does not have to be related to science or technology.

Technical Reports

"Sometimes engineers who are briefing a general have trouble because what interests them, what they think important, or how they think does not match the general or the general's needs."

—MARINE COLONEL

AUDIENCE, SITUATION, PURPOSE

Naval personnel must often work with civilian industry or with reports and other documents prepared by civilian industry. Those doing research at naval labs, naval test centers, or naval schools must help generate technical reports, while others will just have to use them. Specifically, service members will often need to read or prepare technical reports when they work on staffs that monitor military contracts, perhaps while working for an O-5 or O-6 project officer or program manager. That officer may request an analysis, a proposal, or a progress report of some kind. Perhaps the Navy wants to build a helicopter engine that will operate effectively in deserts. The program manager might contract for a report that details what such a design would look like, how reliable the engine would be, and how much it would cost.

In the latter case, the company with the contract would respond with a feasibility report, which Navy officials would use to guide them. The report would first help them determine whether to build the engine. Then, having decided to build it, they would use the report to persuade senior officers and other government officials to award them the funds.

Who are the likely readers of these various reports? The most likely are the technical experts on the program manager's staff. These engineers, technicians, accountants, weapons experts, tacticians, and so forth (some of them military, some civil servants) would receive the task to examine the design, the engine's capability, and the costs in great detail. Possessing the technical background to understand all kinds of charts, tables, diagrams, and descriptions, they would expect detailed technical explanations in these write-ups.

But other readers of the document would not be experts, among them the most important audience—decision makers. These readers would probably not understand all the technical parts of the report, nor would they need the detail that experts require. The project manager, for example, might be generally knowledgeable in tactics and weapon systems but would not necessarily be an engineer. This person would need a semitechnical explanation, one emphasizing conclusions and recommendations. Similarly, senior military and elected officials would also need a semitechnical big-picture discussion of the most important information (rather than a highly technical discussion of all the details).

Thus, a given report may need to include both highly technical and semitechnical content. It is common that one document has to speak clearly both to experts and to decision makers, who tend to be less specialized. With the many other kinds of documents needed for the government to go ahead with a project, each report has to address readers with very different backgrounds and needs. Whether it's a formal proposal by a firm to build the engine, a progress report submitted by the contracted firm, a research report used in the technical design, an instruction manual designed for operators, or a final report submitted upon the project's completion, the document must address multiple audiences.

While we won't cover each kind of report mentioned above, you can refer to one or more of the references on technical writing mentioned at the end of this chapter and in appendix A. We will, however, discuss techniques of structure as well as style principles that industrial writers have used for decades to satisfy such diverse audiences.

ORGANIZATION OF TECHNICAL REPORTS

Following are several proven methods of designing technical reports so that they reach their multiple audiences with the right kind of information, at the necessary reading level, and with the appropriate detail. And as other chapters in this book demonstrate, these methods apply to a variety of complex naval documents, not just technical reports.

Present Conclusions before Rationales

Of course, when investigators have a problem to solve, they typically begin by assessing the problem, then conduct the investigation or research, and eventually come to conclusions. Sometimes investigators write up their reports following this same order, with conclusions last. You may have learned to write academic reports for physics or chemistry labs this way.

Yet as we've seen in other contexts, the chronological order of the investigation may be opposite the reader's needs. Just as naval readers habitually glance at the action paragraph before reading the whole document through (sometimes instead of reading it through at all), decision makers typically look for the conclusions and recommendations first. Often they only skim the rest of the report—they just don't need to read (or don't *think* they need to read) all of it.

FIGURE 8.1. *Organization Comparison between Academic Reports and Technical Reports*

Academic Report	Technical Report
Statement of Problem	Statement of Problem
Procedures	Conclusions and Recommendations
Methodology	Support
Analysis	Methodology
Results	Analysis
Discussion of Results	Results
Conclusions	Discussion of Results
Recommendations	

Help these readers out. As a former board member of the Center for Plain Language points out, "tell readers where you are going before you go there; put your main point before the background and you'll be far less likely to lose them." Figure 8.1 provides an example of what this guidance might mean for the sequence of a standard investigative or research report.

Sometimes you might have to write your report in a rigidly specified format, which may resemble the one in the left column of figure 8.1 (though even this report will usually begin with an executive summary that includes brief conclusions and recommendations). If not, put the conclusions and recommendations at the beginning to give readers a head start on the vital information. As discussed in chapter 1, plain language goes beyond word choice; it includes reader-friendly structure as well.

Subdivide into Short Sections

You can help the reader immensely by organizing your reports into sections. Use care in carving sections out—design them and their headings with the readers' needs in mind.

On hardcopy documents, add paper or plastic tabs to make sections readily visible. For very long documents, consider adding an index, too. With electronic documents, provide internal links for the longer report sections.

Use Reviews, Surveys, and Summaries

Summaries aid everyone. Remember that even those few reviewers who read a report straight through will seldom be able to avoid all interruptions from meetings, office visits, and phone calls while they read. Actually, in the Navy and Marine Corps (as in the corporate world), you are almost always writing for a distracted reader who has many other demands on her or his time. Hence, at strategic locations in your text (such as the beginnings or ends of major sections), review topics that you've presented before and summarize the information that follows after. Above all, after you've written the whole report, from introduction to the conclusions and recommendations, proofread it not simply for grammar and punctuation but also for coherence and unity.

Use Appendices

Relegate to appendices material that only specialists need. Don't let numbers, complicated designs, and other data overburden the text of the report. You should include only some of these elements. Full-length reports, like operation orders, often have many appendices.

Executive Summaries

> *"Most executive summaries are very disappointing; writers seem unable to recognize the key ideas in their own reports. They allocate space in the executive summary in the same proportion as in the original document, whereas they ought to focus only on the essential items, the results. Always think to yourself, 'I'm likely to lose my readers unless I get their attention here.'"*
>
> —PROFESSOR, NAVAL WAR COLLEGE

AUDIENCE AND PURPOSE

The executive summary is an important structural element of technical reports and many other documents. It is a preliminary overview of vital information. Design this section so that by reading it, an executive or other decision maker will have enough information to proceed to a decision. Actually, whoever picks up a report with a good executive summary can use it to get a quick feel for its contents; almost all readers begin with the executive summary if one is provided. Still, its main purpose is to provide executives with what they need to make decisions.

The length of an executive summary varies. It may run twenty pages for a book-length document but only a page or two for a twenty-page report. A good rule of thumb is to make this section no more than one-tenth the size of the report it summarizes. But even more common is to keep it to one page. For example, the highly readable 250-page Columbia Accident Investigative Board Report on the *Columbia* Space Shuttle disaster of 2003 contains only a one-page executive summary.

Place an executive summary at the beginning of a report soon after the title page. Set off the summary from the rest of the document because everyone is likely to read this section since it identifies the gist of the report and its overall import. A tab in a hardcopy report could help make it easy to locate.

You've probably seen executive summaries before, even if they were not labeled as such. An email that introduces a lengthy attachment can fill this role, telling you the point of the document so you'll know whether to open it and read further. A briefing memo for a correspondence package on a staff serves as an executive summary because it also enables a superior to understand the matter at hand quickly without having to page through the whole package. Submarine patrol reports typically begin with executive summaries because not everyone needs all the details. When a personnel board convenes in Millington, Tennessee—the Retired Personnel Board, for example—an executive summary begins the board's report.

Often, when a naval command issues a change to a major instruction, authorities will draw attention to the essence of that change by sending an executive summary in message format.

You can construct the executive summary in various ways. An emphasis on results, conclusions, and recommendations sets it apart from other summaries, such as abstracts, discussed in the next section. The executive summary may vary with situation and audience in how much they discuss background, procedures, and methodology (some treat each of these items briefly, while others don't discuss them at all). The best of them focus mostly on what all the factors lead to, that is,

- what results show,
- what conclusions you can draw, and
- what action your audience should take.

Don't try to cover everything in an executive summary; include only the most essential information. Also, remember to wait to write the executive summary *after* you have written the whole report in its final form; otherwise it may not do justice to or match the actual report. An example of an executive summary appears as the opening section of each chapter of this book.

Abstracts

AUDIENCE AND PURPOSE

Another kind of summary device for technical reports is the abstract, which may preface a report regardless of whether it also has an executive summary. The abstract differs from an executive summary primarily in function and audience. Whereas an executive summary is a synopsis of a report's conclusions and recommendations and is meant to guide decision makers or to prepare other readers for the direction of the main text, an abstract is a screening tool that is usually intended for researchers or other specialists, helping them decide whether to read the report at all.

Writers compose abstracts for professional articles as well as technical reports. But whatever documents they summarize, abstracts have certain standard features:

- They often include great technical detail in a very condensed and highly technical discussion that a layperson will have trouble following.
- They are typically short, from a couple of sentences to about three hundred words (but they still use full sentences, not fragments or bullets).
- They usually are written as one paragraph.
- They don't focus on conclusions or recommendations but either give equal value to every part of a report or concentrate on a project's results, quickly letting the expert reader see the scientific or technical significance of the research.

Write each abstract so it makes sense as a separate document. You have two different styles to choose from.

INFORMATIVE ABSTRACT

The informative abstract is meant to reproduce the report in brief, mirroring all its essential features. In fact, some texts recommend that to write such an abstract, you should first identify the topic sentence from every major section in the report and then simply string all of them together, just smoothing out the wording. You could also work from an outline. Regardless of how you proceed, include in the informative abstract a brief discussion of the background of the research project, its intent, the way you set it up, the procedure you used to carry it out, and the results.

Reading such a summary will tell the researcher whether to order the whole report or not. The following example of an informative abstract is part of a report by a Naval Academy researcher, Asst. Prof. Stephen M. Graham, from August 2003 titled "Fracture Toughness Characterization of HSLA-100 Steels for Carrier Crack Arrestor Applications":

HSLA-100 steel is being considered as a replacement for HY-100 in aircraft carrier crack arrestor applications. The various compositions of HSLA-100 were evaluated and compared to 1.25 in. thick HY-100. Tests were conducted to measure tensile properties, Charpy impact energy, dynamic tear energy, fracture toughness and the reference temperature. The two alloys compared favorably on all tests except the fracture toughness tests at −40°F. HSLA-100 in the T-L orientation exhibited fracture by cleavage after ductile crack growth, whereas the HY-100 remained ductile. This result was unexpected since it is commonly believed that fracture behavior can be correlated with impact tests and the reference temperature. At –20°, fracture remained ductile in the HSLA-100. Consequently, it is recommended that HSLA-100 in the T-L orientation only be used where the minimum service temperature is above –20°F.

Clearly, the emphasis here is not on conclusions and recommendations but rather on faithfully representing the whole report. By using this abstract, researchers looking for information on HSLA-100 steel, fracture-toughness tests, or steel performance at low temperatures might find enough information to decide whether to read the whole report.

The following description of a safety intervention carried out at a Navy mail center, "Ergonomics Intervention at COMNAVREG SW San Diego Mail Center Prevents Injuries" (found in a 2006 Naval Safety Center's online listing of "1,001 Safety Success Stories"), is another good example of an informative abstract. (Although the Naval Safety Center terms it as an "executive summary," it actually serves as an abstract.)

> A routine industrial hygiene survey identified several physical risk factors at the Commander Navy Region Southwest (COMNAVREG SW) San Diego Dockside Mail Center. Heavy lifting and working in awkward postures while processing the large volume of mail handled at NAVSTA San Diego Dockside Mail used to put its mail handlers at risk for work-related musculoskeletal disorders (WMSDs). Funding was provided through the Navy's Hazard Abatement and Mishap Prevention Program (HAMPP) to revamp the mail room service area and purchase ergonomically designed equipment. The estimated savings to the Navy are $41,433.00 every year for a return on investment in 519 days, or approximately one year and five months.

DESCRIPTIVE ABSTRACT

A descriptive abstract is more like a table of contents written in paragraph form. It simply describes from an outside point of view what the report contains. It serves the same general purpose as the informative abstract, but rather than reproducing the original report in brief, the descriptive abstract describes what the report contains.

An example of a descriptive abstract comes from an interim report that the Naval Postgraduate School at Monterey issued in 1987. The report, by Scott A.

Sandgathe, is titled "Opportunities for Tropical Cyclone Motion Research in the Northwest Pacific Region":

> Tropical cyclone track prediction problems in the Northwest Pacific region that need to be researched are reviewed from the perspective of the operational forecaster. This information is provided as background for the upcoming Office of Naval Research field experiment on tropical cyclone motion. A short-term climatology of the frequency and spatial distribution of tropical cyclones is provided. Seven classes of operationally interesting track forecast situations are described. Each cyclone from 1982 through 1985 is tabulated in terms of these classes.

Government and industry widely use both descriptive and informative abstracts. The informative abstract is generally more helpful, for it gives a reader more information. But some kinds of research don't lend themselves easily to informative abstracts. As with an executive summary, compose the abstract last—*after* you have completed the final draft of your report. This way you'll be sure you summarize only what is actually in the report. As you draft the summary, be careful not to add fresh insights or new information that does not appear in the main text.

ABSTRACTS FOR ARTICLES

A different type of abstract is that written for an article. Many technical journals require authors to submit abstracts along with their articles for publication. These are used in abstract databases but often preface the articles themselves.

The following is an example of an abstract published in the PubMed online database that prefaced the article "Viral Gastroenteritis: The USS *Theodore Roosevelt* Experience" in the journal *Military Medicine*. The short piece focuses on the Medical Department's conclusions about the experience they went through. Half descriptive abstract and half executive summary, it typifies abstracts that preface articles in military journals.

> Although the spread of disease on board Navy ships is not a novel concept, the medical department of the USS THEODORE ROOSEVELT recently experienced a significant outbreak of viral gastroenteritis while at sea. The impact on the crew and medical department is reviewed in this case report. The use of the Navy Disease Non-Battle Injury tracking system was validated. Furthermore, we proposed the placement of waterless, isopropyl alcohol-based hand-cleaning systems in strategic locations throughout the

ship to help prevent and minimize the spread of future disease. Finally, more stringent recommendations regarding sick-in-quarters status and careful utilization of consumable resources are necessary components of an effective outbreak management strategy.

KEY WORDS, ABSTRACTS, AND COMPUTERS

Where will someone read this kind of abstract? In an electronic search for published material on a given topic—"missile detector radar," for example—a database (of a library or another repository of specific studies) will usually present the titles and abstracts of all the recent articles. By reading each abstract, the researcher can decide whether to order the complete texts. In many cases the articles themselves will also be accessible online.

As the writer of a military report, if you want to supply your abstract to be on record, you will use the DoD "Report Documentation Page" (Standard Form 298). You'll not only supply an abstract but will also fill out a section called "Key Words." List there the terms that best indicate the substance of your report, both the subjects that it addresses directly and others that it touches on significantly. Take some care in selecting terms; don't make them too complex. In some cases you'll have to select from a predetermined list of standard terms.

For example, the documentation page for the article on tropical-cyclone-motion research mentioned earlier lists "tropical cyclone motion," "tropical meteorology," "tropical cyclone path prediction," and "typhoon motion" as its key words.

Technical Writing Style Tips

"Many writers don't realize that when you use specialized terms in a piece of writing, not only can that make your meaning harder to decipher, it also gives the whole thing a style and tone of being for insiders, and your outside readers may feel as if you don't actually care about them at all."

—LIEUTENANT COMMANDER (RET.), SUPPLY CORPS

PASSIVE VS. ACTIVE VOICE

You may have noticed that the abstract on tropical cyclone motion is written in the passive voice. Indeed, in that abstract every sentence is passive, the main verbs being "are reviewed," "is provided," "is provided," "are described," and "is tabulated." Because principles of plain language insist that you minimize such usage as a general rule, such heavy dependence on the passive calls for comment.

Passive voice is widespread in scientific writing (which is not the same thing as technical writing). Instead of stating, "the chemist observed the experiment," scientists will usually write, "the experiment was observed," not only writing in the passive voice but also omitting all mention of the person doing the observation. The rationale is that normally the results of the scientific investigation are much more important than the investigator, thus the passive voice emphasizes those results.

This usage does focus attention on the receiver of a sentence's action. Often the object of the action is more important than the actor and deserves more attention. To say the "feasibility of ocean surveillance platforms in detecting submarines has repeatedly been demonstrated" focuses attention on that feasibility rather than on who demonstrated it. In some situations that focus may be perfectly appropriate, depending on your purpose.

Some may also think that the passive voice is somehow more objective. This view is incorrect. A passive statement is not inherently more objective than a statement identifying who did the action. For example, changing "*we concluded* that low flying aircraft would not cause mines to detonate" to read "*it was concluded* that low-flying aircraft would not cause mines to detonate" is not more objective. Either way, we know someone drew the conclusion.

Should you rigorously revise the passive voice to active voice in technical documents you must review or sign off on? In most cases, yes, active voice will make them more readable. Yet it depends on the context. In circumstances where such rewriting won't make much difference, or where the revision will cause friction with your boss (for example, if your boss was trained to use the passive voice and deems doing otherwise unprofessional), the improvement probably will not be worth the effort. But wherever passive voice hinders the readability of a text or when you anticipate that the document will have high visibility, you should at least put the executive summary into mostly active voice and perhaps go on to rework the "Conclusions" and "Recommendations" sections as well. See chapter 1 for more discussion about passive and active voice.

NOUN STRINGS

Another major problem plaguing technical writing is the wide use of noun stacks or noun strings. English grammar allows us to put nouns together, with each one becoming an adjective for the one next to it. But the fact that these phrases are allowed does not mean they are always helpful. Noun strings longer than three or four words, such as "aircraft carrier crack arrestor applications" (from one of the abstracts above), "Commander Navy Region Southwest San Diego Dockside Mail Center" (from another one), or "naval writing passive voice abuse advice" (fictitious) are common. Although some writers are accustomed to the use of noun strings, these ultradense phrases can make a document almost unreadable.

Keep an eye out for this tendency, and if you find many noun strings in your own writing, consider breaking up some of them to read more the way you would say them in conversation ("applications for crack arrestors in aircraft carriers," for example, or "Commander of the San Diego Dockside Mail Center in Navy Region Southwest").

For more on these issues and other tips for writing in a clear, audience-friendly style, see the section on plain language in chapter 1.

OTHER TECHNICAL DOCUMENTS

Naval personnel will have to write many other technical documents from time to time. Much of the writing in industry and business is similar to the writing in the naval services—letters, memos, and directives prevail in almost all organizations. While formats and styles differ from one organization to another, a writer can usually recognize a letter anywhere and adapt to the required format quickly.

But some technical documents differ greatly from standard naval correspondence or staff work. Such documents can be very troublesome for those naval personnel who have to compose them. Supply officers and others, for example, often have to write "specifications" for contracts. Also involved in the contracting process are "statements of work," which shipboard officers frequently have to submit. While on staffs, many naval personnel have to draft or revise "position descriptions" and then try to get those positions funded.

You may find the office or lab you work with has put out a style guide to help you. If not, seek guidance for writing such documents from knowledgeable professionals or from standard technical writing texts and sourcebooks (such as those listed below and in appendix A).

EMERGING TECHNOLOGY IN WRITING

USING AI TO WRITE

In recent years applications of artificial intelligence (AI) to produce texts have become more widespread, raising many questions about the future of writing. The use of AI to create (or generate) text goes far beyond the field of writing. ChatGPT and similar tools can help analyze data, compose music, and write computer code. Similarly, applications such as DALL-E simulate the creation of visual arts, raising questions both practical and philosophical about the nature of creativity, and ultimately what it means to be human.

In the world of technical writing—or more broadly the professional naval writing focused on throughout this book—questions of authentic creativity may be less in the foreground than they would be for painting a landscape or writing a novel. Yet the Navy and the other armed services must weigh factors of security and reliability along with originality and the presence of a human operator. As of 2023, ChatGPT and other AI have caused much stir—both enthusiastic and cautious. But because the technology and its application are quickly evolving, what will be considered ethical, legal, and professional in this regard remains to be determined. There seems to be consensus about one thing, however: AI-type writing tools like ChatGPT will grow, not diminish, in the breadth of their use.

Even more pertinent to the focus of this book, below are four recommended best practices regarding the use of AI technology (if eventually allowed at all) to generate reports, fitreps, correspondence, policy, or any other form of Navy or Marine Corps writing.

Consider currency. Ensure the AI technology draws on current and relevant source material. In 2023, for example, ChatGPT was not accessing content written after 2021.

Review and revise. Whatever type of document (policy, eval, memo, report, or any other) you use AI to generate, do not accept the output as a finished product without careful review and revision. Just as a human operator should remain part of any autonomous weapon use, a human writer should carefully review and revise as needed any text generated by AI to ensure it is well tailored in content, structure, and style for the intended audience and purpose. If you are permitted to use text-generating AI, do not assume the output provides the nuance required for the audience and situation just because it sounds generally professional. Through a rigorous revision process, make the text your own.

Monitor legality. In 2023, lawsuits over art or music generated by AI challenged its originality and demanded credit to the source data the applications have used.

Ensure security. A major concern for defense applications of such AI is whether it represents a breach of security for it to draw on and repurpose technical data and text samples needed for generating a document. Research and follow security guidelines (which may be in development for some years to come) before using these tools.

References on Technical Writing

Two helpful civilian textbooks on technical writing are Mike Markel, *Technical Communication*, 13th edition (St. Martin's, 2020); and John M. Lannon and Laura J. Gurak, *Technical Communication*, 15th edition (Pearson, 2019). Both cover the basic kinds of technical documents (such as reports, proposals, and progress reports, among others) as well as abstracts, executive summaries, technical illustrations, and so on. Each has gone through several editions.

Another useful text is Charles T. Brusaw, Gerald J. Alred, and Walter E. Oliu, *Handbook of Technical Writing*, 12th edition (St. Martin's, 2020), which serves as a dictionary of technical-writing terms and concepts.

Two helpful guides to technical editing are Anne Eisenberg, *Guide to Technical Editing: Discussion, Dictionary, and Exercises* (Oxford University Press, 1992); and *The Microsoft Manual of Style for Technical Publications*, 4th edition (Microsoft, 2012). Eisenberg's publication, though older than the other guides, is a practical workbook, filled with examples and exercises.

Beyond these, many naval or other military offices have their own research guides, which can be helpful. Some are oriented specifically toward historical research, while others are more technically or operationally focused. For example, the Marine Corps Historical Center and the Naval War College publish helpful internal guides.

An excellent online source for researching military articles is the Air University Library Index of Military Periodicals (AULIMP), which includes naval as well as general military journals. Also available online is the Staff College Automated Military Periodical Index (SCAMPI), an index of military periodicals hosted by the Joint Forces Staff College.

Finally, one should become familiar with the resources used widely by naval and defense researchers, specifically the abstracting services put out by the Department of Commerce's National Technical Information Service (NTIS), the Defense Technical Information Center (DTIC), and (for logistics) the Defense Logistics Agency (DLA). Depending on the topic, there are literally hundreds of abstracting services upon which naval researchers might call. Librarians and subject-matter experts are proficient in this kind of data retrieval.

9

Report of a JAGMAN Investigation

> *"I was hit with a JAGMAN two weeks after reporting as a brand-new second lieutenant. A JAGMAN is a 'rite of passage,' assigned not on the basis of who's best qualified to do it—but on the basis of who's available."*
> —MARINE CAPTAIN, INSTRUCTOR AT U.S. NAVAL ACADEMY

EXECUTIVE SUMMARY

A JAGMAN investigation and the ensuing report produce a common management tool in helping the command determine the causes of a mishap or a violation of military law. While the report does not have the legal weight of a criminal investigation in a court of law, it can help the CO determine a direction of further action. Conducting an investigation and writing the report require careful planning and thoughtful steps to gather evidence and to present an objective and clear analysis of the evidence. You will assemble the elements of the report in a specific order. Your thoroughness and clarity will not only reflect on your professionalism but also greatly help the command make the best decision about the matter in question.

WHAT IS A JAGMAN INVESTIGATION?

A *Manual of the Judge Advocate General* (JAG Manual, or JAGMAN) investigation—and the report that expresses its findings—is a management tool, a means by which a command can gather the facts needed to make a decision surrounding some potential violation of military law. That decision might concern public or congressional inquiries, loss or compromise of classified material, claims for or against the government, destruction of property, accidents, injuries to personnel, or loss of life. Even minor investigations can have substantial ramifications. JOs and senior enlisted are often assigned to conduct the investigations.

When assigned an investigation, you gather information, express considered opinions, and make careful recommendations that will help a commander (the "convening authority") make intelligent decisions. In some cases people's careers depend on what appears in a JAGMAN investigation; findings routinely affect medical retirement, disability pay, veteran's benefits, and promotion opportunities. In others, many thousands of dollars are at issue. As you can see, doing such an investigation is an important responsibility. Conducting one can also teach a person a good deal about command decision making.

The type of JAGMAN investigation covered in this chapter is the "command investigation." It is the most common type and the one most often assigned to someone who has never done one before.

Another kind of investigation you might have to do is the "litigation-report investigation," which differs from the command investigation in that it targets incidents involving potential claims for or against the government. In many respects this type of investigation (designed partly to help protect government-privileged information in civil proceedings) resembles a command investigation except that the investigating officer works directly under the supervision of a JAG officer and pays special attention to the rules involving claims. If you can do a command investigation, you'll be able to do the litigation one, too (especially since you'll personally be given special instructions by a judge advocate). So in this chapter we limit our discussion mainly to the command investigation and refer to it as a "JAGMAN investigation," which continues to be its informal name.

One other brief legal survey you might be called upon to conduct is the "preliminary inquiry." This is an initial search into an incident designed to discover whether any investigation should be conducted and if so, what kind. A preliminary inquiry of some sort (though often an extremely informal one) is supposed to precede most JAGMAN investigations. A brief discussion of these inquiries (with a sample write-up) is at the end of this chapter.

THE JAGMAN INVESTIGATION IN GENERAL

If you are assigned to conduct a JAGMAN investigation, understand several important points. First, a JAGMAN investigation will normally become your primary duty until you complete it. Second, doing an investigation is an excellent opportunity to get the attention of your commander because the CO will review and personally endorse your report once you complete it. The CO can ask you to do it over if it isn't as good as he or she thinks it should be. (On the other hand, as investigating officer you can recommend that the CO enlarge, restrict, or modify the scope of the investigation or change any instruction in the appointing order.) Third, realize that this investigation will often go up the chain of command, through your boss's boss and maybe even higher, up to the Office of the Judge Advocate General. Sometimes the CO will decide the investigation doesn't need to be forwarded—but often it will be.

Should anyone have criticisms along the way, either of the report you prepare or of your command's way of doing business as reflected in the report, they won't be private. Your boss will hear of them for sure and possibly many other officers as well. Those critiques could reflect both on you and on your ship or unit.

These investigations are not kept within the military, either. Because of the Freedom of Information Act, they are not proprietary information, and virtually anyone can get copies. Unlike litigation-report investigations, which are kept privileged, command investigations are routinely distributed; the JAG Office sends out thousands of copies a year. Next of kin, lawyers, reporters, members of Congress—all can request a copy of a report (and can almost always get it). In addition, a JAGMAN investigation involving the death of a service member goes to the next of kin as a matter of course.

These investigations can have higher visibility than anything else a JO or senior enlisted person will write. Moreover, unlike other important documents that a junior person will usually only draft but not sign, the person conducting the JAGMAN investigation also signs the report. All of these factors—to say nothing of the most important matter, seeing that justice and truth are served, both for the service member and for the Navy or Marine Corps itself—suggest you should do your very best with this demanding research-and-writing task.

What follows is a brief overview on conducting a JAGMAN investigation and writing the report. This will serve as a starting point. Of course, do not depend on this chapter alone, but rely heavily on the official sources, especially the JAG Manual. Also consult official instructions, chain-of-command directives, and the other helpful guides available to you. As you'll see, you have many sources at hand to help you both conduct and write up this kind of investigation.

REFERENCES ON JAGMAN INVESTIGATIONS

- Article 31, Uniform Code of Military Justice (UCMJ).
- *Manual of the Judge Advocate General* (JAG Manual, or JAGMAN). This is, of course, the standard reference on JAGMAN investigations, as on many other things.
- Of special importance for JAGMAN investigations are the sections in the JAG Manual on the following:
  - Privacy Act compliance

- · The Appointing Order
- · The Investigative Report itself
- · Line-of-Duty (LOD) and Misconduct Determinations
- · Injury/Disease Warnings
- · Article 31 Warnings
- · Investigations of Specific Types of Incidents
- · Checklists of Various Kinds
- · Claims for or against the Government
- *JAGMAN Investigations Handbook*, published by the Naval Justice School (NJS). Also referred to as the "NJS Handbook," it provides detailed guidance on investigations, including checklists and sample documents of many kinds.
- *Security Manual*, OPNAVINST 5510.1 series. See the sections and exhibits discussing JAGMAN investigations.
- MILPERSMAN, Section 4210100. Consult for death cases.
- Checklists and instructions issued by the local JAG officials and various type and administrative commanders.

Preliminary Inquiry

We open this discussion with a brief section about a legal report you may have to do prior to a JAGMAN investigation. Preceding any formal investigation, a command will often conduct a preliminary inquiry. This is a preliminary and informal survey, basically a "quick look" at an incident, designed to determine whether any official investigation ought to be conducted at all and if so, what kind. In this inquiry a CO may choose any means at all to look into an incident, may conduct this search personally or delegate it to someone else, and may determine whether to document these actions in writing.

If she or he decides to assign it to you, the CO will direct that you complete the inquiry within three days. If you find you can't finish it within that time, you might be trying to do too much and should go back to your CO for further guidance. When assigned a preliminary inquiry, begin immediately, using (if available) the Preliminary Investigation Checklist from the *JAGMAN Investigations Handbook*. Check first to see if the incident is under investigation by any civilian or other military agencies and to determine whether it should be considered a "major" incident. In either of these cases, follow the respective guidelines in chapter 2 of the JAGMAN, otherwise

- obtain available documentation pertaining to the inquiry;

- locate and preserve evidence;

- draw up a list of possible witnesses; and

- interview those witnesses in person, by phone, or by message, advising each of their rights and obtaining Privacy Act Statements as necessary (all of this as outlined earlier in this chapter).

Pursue all of the above just so far as necessary to make an informed recommendation as to what specific course the convening authority should pursue, that is,

- no further action,

- a command investigation (normal JAGMAN),

- a litigation-report investigation, or

- a recommendation to convene a court or board of inquiry.

After following these steps, report back to the convening authority with your recommendation, meanwhile preserving all documentation, evidence, and witness statements for any eventual investigation.

FIGURE 9.1. *Preliminary Inquiry*

10 Nov. 2020

From:　LTJG Paul Robichaux, USNR, 1115

To:　　Commanding Officer, USS WHIDBEY ISLAND (LSD 41)

Subj:　PRELIMINARY INQUIRY INTO THE RUMORED COLLISION OF LCM 8-1 AND PRIVATELY OWNED FISHING VESSELS AT MOREHEAD CITY, NC, ON 8 NOVEMBER 2020

Ref:　　JAGMAN Section 0204

1. On the evening of 8 November 2020, WHIDBEY ISLAND personnel overheard rumors that charges were to be made against the ship because of an alleged collision of one of the ship's LCMs with some fishing boats that supposedly had occurred that morning. I was assigned to conduct a preliminary investigation into this rumored incident. This is the report of my investigation.

2. Personnel contacted:
   a. BM2 R. F. DOUGHERTY, USN, USS WHIDBEY ISLAND (LSD 41). Dougherty was the Boat Officer aboard LCM 8-1 at the time of the incident.
   b. BM2 H. D. SHENK, USN, USS WHIDBEY ISLAND (LSD 41). Shenk was the Boat Coxswain aboard LCM 8-1 at the time of the incident.
   c. Mr. C. O. Smith, owner of a fishing boat moored at the Morehead City docks, telephone 804-111-1111. Mr. Smith was a witness to the incident.

3. Materials reviewed:
   a. Bow and port side of LCM 8-1.
   b. Fishing boats "Jennifer," "Faring Well," and "Jennifer II"
   c. Pilings nearby these fishing boats.

FIGURE 9.1. *(continued)*

4. Summary of findings:

   a. At about 0900 on 8 November, LCM 8-1 was towing another assault craft in the vicinity of private boat docks at Morehead City, NC. The port engine became fouled with a tow rope, and strong winds and current made it difficult to maneuver the craft. The lead LCM came into brief contact with dock pilings near the privately owned fishing craft named in paragraph 3(b).

   b. BM2 Daugherty and BM2 Shenk report that LCM 8-1 did not contact or damage any of the fishing boats. Mr. C. O. Smith (who was working on his boat at the time and who witnessed the event) agreed that although the lead LCM came as close as ten feet to one of the fishing boats, no contact actually occurred. Moreover, he reports that two of the fishing boats mentioned above ("Jennifer" and "Faring Well") were taken to sea soon afterward. The third ("Jennifer II") was moored well away from the pilings, too far away to have been involved.

   c. My observation of the fishing boats from the dock indicated no obvious recent damage. No gray paint from LCM 8-1 or other indication of a possible collision was visible.

   d. The owner of the fishing boats mentioned above is Mr. J. R. Turner, telephone 804-222-222. I did not speak to Mr. Turner.

5. Recommendation:

   That no action be taken at this time. If someone makes a formal complaint, I recommend Commanding Officer, USS WHIDBEY ISLAND, immediately initiate a Litigation-Report Investigation of this incident.

Paul Robichaux, LTJG, USNR

Although written documentation of the preliminary inquiry is not mandatory, your boss may ask you to write up your findings. If so, follow the format in the JAGMAN. The fictional example in figure 9.1 follows that format and is based upon an actual inquiry.

First Steps of a JAGMAN Investigation

> *"I approach each investigation with this attitude:*
> *I don't know what happened—this investigation is supposed to explain it to me."*
>
> —JAG LIEUTENANT COMMANDER

Some officers in the fleet and field say that the problem with a JAGMAN investigation is not writing the report but doing the investigating. They argue that the tough part is digging down to the underlying, determining facts of the case. Yet those who review JAGMAN investigations insist (from having read dozens of reports) that the writing is a big problem, too. So we'll look at both tasks. We'll start by discussing what you should do to begin your investigation.

1. Read the whole section on JAGMAN investigations in this book for general familiarity. As you do, pay special attention to the various terms and concepts involved.

2. Read the appointing order very carefully. It is your basic marching order, and it should address your specific investigation. If you don't understand any item in the appointing order, ask the "convening authority" (which is usually your CO) about it. Further, if during your investigation you feel that you should broaden or narrow the scope of the inquiry or that you need to change any instructions in the appointing order, submit a request (orally or in writing) to the convening authority. Realize, too, that any single JAGMAN investigation should cover only one incident. If you find you're really investigating two or three separate incidents, report this early on.

3. Get hold of an updated JAG Manual and review the sections mentioned specifically in the appointing order and those listed under "References on JAGMAN Investigations" above, especially those sections involving your particular case. For instance, if you must make an LOD/misconduct determination, be sure to review that section of the JAGMAN; if you are investigating a death, review the section on death cases; and so on.

4. Review recently completed JAGMAN investigations from files at your ship or station or the samples and guidance in the NJS Handbook. Reading a few will give you more familiarity with what they look like and, perhaps more important, what they customarily look into. (The examples later in this chapter will provide a start.)

5. Talk to an officer (or chief) on board your ship or station who has done investigations before or who has been to a naval school on this or related subjects. This individual should be able to give you both practical advice and good written guidance. Yet if this person is also the ship's or station's legal officer, he or she might have to keep a distance from the case. If it goes to mast or trial, this person may be required to get involved in the case later. Beyond this situation, the legal officer will usually have the duty of drafting the first endorsement for the CO.

6. Find out where the local staff JAG is or, if that officer is not available locally, seek out another naval lawyer (perhaps at a local Naval Legal Services Office) who might be able to give you some help. Most staff JAGs are more than willing to discuss with you how to go about an investigation, and later they will be glad to look at your rough draft. Indeed, some staff JAGs expect you to visit them first and then to send them a rough draft before you formally submit your report. Such a review will often save them time later. (Of course, don't communicate with the staff JAG unless your CO first gives consent.)

7. Don't forget to take counsel from those in your own chain of command, such as the XO. JAG officers are not necessarily attuned to the special sensitivities of operational commands or to all the intricacies of shipboard situations. Whomever you talk to, make sure you get a good feeling for

   · where to go for information;
   · what to look for;
   · who to talk to;
   · what kind and number of questions to ask;
   · any available checklists concerning the specific kind of incident you are investigating (beyond those discussed in the section "Conducting the Investigation" below); and
   · when to stop investigating and start writing.

Elements of the Investigation

PRIVACY ACT STATEMENTS

As the investigating officer, you must ask individuals to sign Privacy Act Statements whenever you request them to disclose private (personal) information about themselves. Such situations, however, are not the rule, and you should avoid unnecessary use of such statements.

During an investigation into a loss of funds, for example, if you ask an accountable individual to disclose his or her personal financial status, this disclosure is subject to the Privacy Act. But asking a service member to account for actions when on watch, to recount actions in the course of official duties, or to relate events observed in the course of routine activities is not a request for private information and does not require a privacy warning.

The requirement for a Privacy Act Statement is spelled out in the JAGMAN, where you will find a format for one in an appendix. Very good formats also appear in an appendix to the NJS Handbook and in many local instructions.

INJURY/DISEASE WARNINGS

Don't ask service members about the origin or aggravation of any disease, injury, or disability they have suffered without first advising them of their statutory rights not to make such a statement. Have them sign JAGMAN warnings that they have read and understood their rights before proceeding with interviews. Again, a proper warning form is in an appendix of the NJS Handbook.

ARTICLE 31 WARNING

Whenever a person is suspected of committing an offense under the UCMJ, you must advise that person of his or her rights under Article 31, UCMJ, before proceeding with any questions. These so-called Article 31 Rights include the right to remain silent, the right to consult with a lawyer, and the right to terminate the interview at any time as well as the warning that any statements made might be used against that person in trial by court-martial. (A proper warning form is in an appendix of the JAG Manual and also in the NJS Handbook.)

LINE-OF-DUTY/MISCONDUCT DETERMINATIONS

Perhaps the most common JAGMAN investigations are cases of injury and disease. Here a major part of your responsibility is to help your CO make a "line-of-duty/misconduct" determination; that is, to help the chain of command determine, first, if an injury incurred "in the line of duty" and, second, whether it involved misconduct—two separate determinations. Many rights and benefits depend on these findings.

What constitutes "line of duty," and what is "misconduct"? LOD does not have to mean you were "on duty" or doing Navy or Marine Corps work when the injury occurred. Put briefly, the service presumes you've incurred any injuries or diseases "in the line of duty" unless clear and convincing evidence exists otherwise. What would such evidence be? If a service member incurred an injury as a result of misconduct, while deserting, or while an unauthorized absentee in excess of twenty-four hours, then the injury might be regarded as "not in the line of duty."

A finding of "misconduct," on the other hand, would come about if an investigation clearly showed that a service member intentionally incurred an injury or that it was "the proximate result of such gross negligence as to demonstrate a reckless disregard of the consequences." These rules are pretty clear, but of course, pinning down a clear determination in individual cases can be tricky.

Specific relationships exist between misconduct and LOD. For example, a determination of misconduct always requires a determination of "not in the line of duty." To put the whole issue simply, the finding must be one of these three:

1. In the line of duty, not due to one's own misconduct.
2. Not in the line of duty, not due to one's own misconduct.
3. Not in the line of duty, due to one's own misconduct.

Consult the JAGMAN section on "Line-of-Duty and Misconduct Determinations." In particular, study the special rules on intoxication, on mental responsibility, and on suicidal acts and gestures to make sure you fully understand these determinations.

You should realize also that a determination of misconduct is not a punitive measure. While it may directly affect such compensations as VA benefits, medical retirement, and disability pay, a favorable or unfavorable determination has no binding power on any issue of guilt or innocence in a disciplinary proceeding. Of course, an investigator has the responsibility to draw up specific charges if findings suggest they are warranted. But the determinations of "line of duty" and "misconduct" are administrative rather than judicial. A JAGMAN investigation is not a legal proceeding or a court of law.

Indeed, once having had an injury investigated, a commander may report the LOD/misconduct determination informally by an entry in a health record, on a form, or in a letter report. The commander does not have to send the JAGMAN investigation up the chain. Many investigators forget this possibility. As a senior judge advocate once remarked:

> My advice to commands/investigating officers is to use the documentation vehicle that requires the least amount of work and that is sufficient to protect the rights of the service member and the interests of the government. I am not suggesting that we look for ways to avoid work. What I mean is that you should use health or dental record entries and forms or letter reports (in accordance with JAGMAN) when appropriate. I see too many JAG Manual Investigations that did not need to be done.

"Always remember—Mom is the ultimate recipient of a death investigation. The family's anguish of death is compounded by your stating in the investigation that 'SN Smith was a scumbag.' There's no reason to say that even if it's true."
—CHIEF LEGALMAN

Line of Duty/Misconduct General Guidelines

The following are guidelines to follow concerning LOD/misconduct determinations. These guidelines are in addition to other guidelines for JAGMAN investigations.

- Ensure you understand fully LOD, misconduct, the relationship between these concepts, and all special rules involved.
- Make a finding of fact as to the leave, liberty, or duty status of any injured person.
- See that the investigation clarifies the nature and extent of all injuries and includes the place, extent, and cause of any hospitalization. Especially ensure that you differentiate periods of alcohol or drug impairment and periods of psychiatric treatment.

- Ensure that you state clearly the amount of lost work time, if any, as a finding of fact. If the injured person is still disabled when you submit the report, include a medical officer's prognosis.

- The convening authority will afford a JAGMAN hearing to any service member who is thought to have been injured or diseased either "not in the line of duty" or "due to own misconduct" and will append the hearing results as an enclosure to the first endorsement to the investigation. Appropriate hearing forms are in the appendixes to the JAGMAN.

- Fill out an LOD/misconduct checklist as you conduct your investigation. See JAGMAN, NJS Handbook, or local directives.

- Finally, place your opinion as to the LOO/misconduct determination in the "Opinions" section of the investigation.

CONDUCTING THE INVESTIGATION

"I remember doing one JAGMAN investigation in my career; I was a lieutenant. Supply reported they had lost a clear plastic canopy for an F-18. How do you lose a canopy? The canopy had cost the Navy $273,000. So I began the investigation, and started a round of interviewing. I was still working on it two days later, when Supply called me up: 'We found it!' It's remarkable how quickly you can find something when they begin to take it out of your pay."

—NAVY CAPTAIN

GET THE RIGHT CHECKLISTS

Assemble all the relevant checklists. Two are standard—the JAGMAN section "Investigative Report," which is an overall guide on doing the investigation, and the JAGMAN "Checklist for Fact-Finding Bodies," which is an LOD/misconduct checklist. There are also many specialized checklists. Several—"Aircraft Accidents," "Vehicle Accidents," "Explosions," "Postal Violations," and others—are found in the JAGMAN section "Investigations of Specific Types of Incidents" and in the NJS Handbook.

There are also locally prepared checklists, some of which deal with events that commonly occur in a particular type of ship or unit, while others cover standard situations even more thoroughly than the JAG Manual. You should ask about these lists as you talk to your XO, local JAG, and other JAG officers. Indeed, sometimes you will find you are required to follow a special checklist due to particulars of your investigation. In any case, local lists may help you greatly in figuring out what to look for and in making sure you've researched all the right data. They'll tell you what witnesses to interview and the kinds of documents to seek out; sometimes they'll even suggest physical evidence to look for.

If a single incident you are investigating involves more than one of the categories specifically listed in the JAG Manual section "Investigations of Specific Types of Incidents," you should go through the checklist for each category involved.

PLAN CAREFULLY BEFORE INTERVIEWING WITNESSES

"Once a lawsuit is filed, it is likely the investigating officer will have been transferred and witnesses will have left the area. It is time consuming, frustrating, and often counterproductive to try to reconstruct an incident or correct a slip-shod investigation after months or years have passed."
—*JAGMAN INVESTIGATIONS HANDBOOK*

Have a plan for going about your interviews. Do them in a reasonable order. Again, ensure you know what warnings to give (see JAGMAN on warnings and Article 31, UCMJ) and be sure to give them before each interview.

You should ordinarily collect relevant information from all other sources prior to interviewing persons suspected of an offense or improper performance of duty, though don't delay questioning those who are likely to transfer or deploy soon. When the interview begins, make sure you give the proper warning (see above) and document your warnings afterward.

Also, for witnesses suspected of an offense or other difficult parties, do your homework. Research the case as much as possible before you see them, write down some questions ahead of time, check with lawyers, and so on. The more complex the case, the more homework you'll want to do before you talk to the key people.

Conducting interviews in person is best, but for witnesses who are out of town or otherwise hard to reach, you can conduct telephone, videoconference, mail, message, or even email interviews (though remember that mail will take a while). For each live interview, prepare a written memorandum for record, setting down the substance of the conversation, the time and date it took place, full identification of the interviewees, and any rights or warnings provided.

Begin an in-person interview by sitting down with the person, giving out a voluntary statement form, and asking the person to write down all relevant facts surrounding the incident. Then ask specific questions orally if you want to make sure you cover particular points. Work any oral responses into a final draft of the person's statement and have the individual review the statement as soon as possible.

When taking a statement from anyone, be sure to phrase it in the actual language of the witness. Try to obtain a signature, but if the witness cannot or will not provide it, certify it yourself to be an accurate summary or the verbatim transcript of oral statements the witness made.

Most importantly, be sure the witnesses speak as factually and specifically as possible. Vague statements such as "pretty drunk," "a few beers," and "pretty fast" do not specify events clearly enough to be very helpful. Try to pin the witness down. For example, instead of accepting "pretty drunk," use a series of questions to obtain a more detailed description:

- How long did you observe this person?
- How clear was his speech?
- Did you observe him walk?
- What was the condition of his eyes?
- What did he smell like?
- What was he drinking?
- Exactly how much had he drunk?
- Over what period of time?

FIND RELEVANT DOCUMENTS AND PHYSICAL EVIDENCE

General

In general, include whatever of the following may be useful as real or documentary evidence: photographs, records, operating logs, directives, watch lists, pieces of damaged equipment, sketches, military or civilian police accident reports, autopsy reports, and hospitalization or clinical records (although watch out for Privacy Act problems), among others. The NJS Handbook contains good guidelines for collecting each kind of evidence. The following are a few guidelines for particular types of cases. See the JAGMAN checklists for additional pointers on your specific case. Also see the section "Documents and Enclosures" below for further guidance on documentary evidence.

Automobile Accidents

In cases involving an automobile accident, include maps, charts, diagrams, or photographs of an accident scene or of a vehicle or other evidence as needed or available. Identify the date and subject matter of any photographs. You can write notes and refer to directions, objects, or elevations and attach those references to the reverse of the photo, if helpful. Note that photos used as enclosures, like the rest of the investigation, are subject to public release under the Freedom of Information Act, so normally don't include gruesome photos of dead bodies or bloody weapons, beds, or floors and the like.

Death

In the case of a death, make sure that the death certificate supports a finding of fact as to the time and cause. But don't wait to submit a death investigation beyond the mandatory processing time if you still haven't obtained the death certificate. Instead, send the investigation on, noting "death certificate to follow." When the certificate comes, send it on to whomever holds the investigation at the time.

LOD/Misconduct

In LOD/misconduct cases, locate and include documentary evidence that substantiates the member's duty status at the time of any injury, disease, or death. This evidence could be an email or message report of the member's duty status, certified copies of service-record documents, or a written statement from the division officer, platoon leader, personnel officer, or other person authorized to grant liberty or record leave status. Note that an unsupported statement by the investigating officer as to the duty status of the individual is not an acceptable substitute for documentary evidence of such status.

Keep JAGMAN Investigations Separate from Other Investigations

More than one kind of investigation may go on with respect to any one incident. Make sure your JAGMAN investigation is completely separate from (and has no reference to) any aircraft-mishap investigation reports, inspector-general reports, or medical quality assurance investigations that are underway. Do not include any reference at all to polygraph examinations. Do not make any use of the narrative of any Naval Criminal Investigative Service (NCIS) investigation that may be going on.

You may, however, use the exhibits of an NCIS investigation in your report. NCIS investigators are usually extremely thorough and professional about giving warnings, questioning, and so on. If you know the NCIS is talking to someone you would also like to talk to (about misconduct, for instance), you might go on with other aspects of your investigation and then ask to see (and use in your report, if pertinent) advance copies of applicable statements by witnesses or similar exhibits. This way you might find that part of your work is already done. Of course, you may then have to conduct additional interviews on your own. On this subject, see JAGMAN sections on "Noncombinable Investigations" and "Investigations Required by Other Regulations."

FOLLOW ADDITIONAL INVESTIGATION GUIDELINES

- Get started quickly. Witnesses will be more likely to be on hand and to have fresh memories, ships or units may still be in the area, and damaged equipment/materials are more apt to be in the same relative position and condition if you get to them quickly.

- Be careful to observe the time limits typically specified—that is, thirty days to complete an investigation from the date of the investigator's appointment.

- Be farseeing and request any delay as soon as you anticipate a need for one. But try your best to get the investigation done quickly. Some investigations stretch over months because of unforeseen ship movements, temporary additional duty (TAD), or other such interruptions. If you are not alert, important witnesses may suddenly turn out to be hundreds of miles away or even out of the country because of transfer or deployment.

- You may acquire evidence in any reasonable manner, and the Military Rules of Evidence required for courts-martial do not bind investigating officers. The reason is that a JAGMAN investigation is purely administrative in nature and not judicial. Its report is strictly advisory, and its opinions are not final determinations or legal judgments. On the other hand, if investigating officers uncover good evidence for use in criminal actions related to the investigation, you may invalidate the evidence if, in acquiring it, you have ignored rules of evidence. So keep your wits about you here, too.

- Remember, the overall purpose of your investigation is to tell a complete story by answering the standard questions who, what, where, when, how, and why. As you go through your investigation, keep brainstorming with these six basic questions in mind until you're sure you've answered them all.

- Realize, of course, that investigations containing sensitive matter must be classified (see OPNAVINST 5510.1 and appropriate sections of the JAGMAN). Remember that if at any time in your work you find you are investigating possible claims for or against the government, you should probably be doing a litigation-report investigation (rather than a command investigation). Immediately inform your CO and be guided by his or her direction.

- Having considered all the above, begin your investigation.

WRITING THE REPORT OF THE INVESTIGATION, SECTION BY SECTION

"The challenge in writing a JAGMAN investigation is to get across what happened. Often the writers leave obvious questions unanswered."

—JAG LIEUTENANT

Because these reports have a stipulated format, the paragraphs below contain a detailed discussion of how to write each section of the JAGMAN investigation. You shouldn't begin writing immediately, however. Instead, first work on grouping the facts. By making use of outlines or note cards, you can work to see the main structures in your material. You can fit all the facts together, separating them from the opinions and recommendations. If you have many facts, you can try arranging them in different patterns.

First focus on the big picture you will present. Then proceed to paint in all the details according to the requirements of each section of the report, as discussed below. Look to the examples for guidance; although all the personal names included are fictional, the examples have been carefully crafted to resemble actual JAGMAN investigations.

We must point out one other thing. When you read JAGMAN investigations and other legal writing, you'll notice that writers use the passive voice widely (alongside other legalisms). Normally there is no need to. As much as possible, write in active voice here as in other naval writing (see chapter 1 for an extended discussion of passive versus active voice).

SUBJECT LINE

Subject lines for JAGMAN investigations differ from other subject lines chiefly in their length; don't be concerned if you need three to five lines to identify the incident thoroughly. Usually, the subject line will be the same one specified in the appointing order—but you may find it needs to be amended. In any case, the subject line should use ALL CAPS, begin with the word "INVESTIGATION," and then go on to cite the following details:

- the basic nature of the incident—that is, an accident, a collision, or other;
- the identity of the unit or ship that was involved in the incident, or of the service member who was involved, or both; and
- the date the incident occurred or was discovered.

Examples of Subject Lines

Here are two good subject lines. Each of them identifies the nature of the incident, the identity of the unit or individual involved, and the date.

> Subject: INVESTIGATION INTO THE CIRCUMSTANCES SURROUND-ING THE SHORTAGE OF FUNDS IN THE SHIP'S STORE ON BOARD USS ZUMWALT (DDG 40) WHICH WAS DISCOVERED IN JANUARY 2023

> Subject: INVESTIGATION TO INQUIRE INTO THE CIRCUMSTANCES CONNECTED WITH THE PHYSICAL INJURY/ACCIDENT THAT OCCURRED ON 13 FEBRUARY 1992 INVOLVING SERGEANT GEORGE F. W—, USMC

LIST OF ENCLOSURES

The JAG Manual requires that the appointing order be the first enclosure to a JAG-MAN investigation, and it should usually be followed by any requests for extension. List subsequent enclosures in the order mentioned in the investigation. See the complete JAGMAN investigation in figure 9.2 for an example of a list of enclosures.

PRELIMINARY STATEMENT

In the preliminary statement you inform the convening authority of the nature of the investigation and of any difficulties you had in complying with the appointing order or in procuring evidence. Also call attention to any other difficulties encountered in this particular investigation. First refer to the appointing order, then comment on as many of the following as are pertinent:

- the general nature of the investigation
- whether you carried out the appointing order and all other directives of the convening authority (some may prove impossible)
- any difficulties encountered, including trouble in gathering information or in ascertaining a particular fact
- conflicts in evidence, if any, and how you resolved them
- reasons for any delay (you should have requested one earlier, and the written request and approval will be enclosures, but still mention the reasons in the preliminary statement)
- the name and organization of any judge advocate consulted
- whether you have advised persons of various rights, as required
- any refusal by a service member to sign or make a statement concerning disease or injury

- any other preparatory information necessary for a complete understanding of the case

DO NOT include a synopsis of the facts here—the findings of facts should tell the basic story. DO NOT include opinions or recommendations. DO NOT include your own itinerary for doing the investigation.

Examples of Preliminary Statements

PRELIMINARY STATEMENT

1. Following reference (a) and enclosure (1), a command investigation was conducted to inquire into the circumstances surrounding injuries sustained by Petty Officer Second Class Jack M. Sweetman in an automobile accident on 25 October 1995.

2. I encountered difficulties obtaining the police report from the Orleans Parish Police Department and the medical bills from Mercy Hospital. An extension was granted to 21 November (Encl (2) and (3)).

3. Although several enclosures indicate that Petty Officer Sweetman had been drinking, the report on his blood alcohol analysis was not available at the time of this report.

Most preliminary statements are shorter; the one below specifically mentions the advising of rights:

PRELIMINARY STATEMENT

1. As directed by enclosure (1) and in accordance with reference (a), an investigation was conducted to inquire into the circumstances surrounding the damage sustained by SH-3H BUNO 152131 (also described in this report by its MODEX number, 519) on board USS KITTY HAWK at or about 2200, 17 March 1984. There were no difficulties encountered in obtaining evidence or information required to complete the investigation; all parties interviewed provided testimony or cooperated fully in all respects. The investigating officer consulted LCDR Jeffrey A. T—, USN, Command Judge Advocate, USS KITTY HAWK, on several occasions.

2. Based upon testimony received, AA John R. J— and AA Earl F. R— were warned that they might be suspected of dereliction of duty and were advised of their rights under Article 31, UCMJ.

FINDINGS OF FACT

"They try to do the JAGMAN too quickly. They shouldn't be thinking of opinions and recommendations before they've got all the Findings of Fact."

—CHIEF LEGALMAN

In the findings-of-fact section, the investigating officer assembles all pertinent evidence. Strive to present the reviewer with an accurate picture of exactly what happened—the specifics as to times, places, and events—in the most logical and clear manner possible.

Although the JAG Manual states that you may group facts into narrative form (that is, paragraphs that tell a story), most JAG officers find this format cumbersome to work with and recommend that you list each fact separately. Once you're sure you have all the facts, assembling the information chronologically is usually best; normally, a chronological order is more coherent for the reader.

Admittedly, more than one series of events may have been taking place simultaneously, and you'll have to adjust by first narrating one series of events, then going back chronologically to describe another. Don't jump around aimlessly, leaving facts in haphazard order, or you will confuse the reader as to exactly what did happen. Do your best to be as coherent as possible. As an O-5 JAG on a major type command's staff remarked: "I can tell a good investigation if, after reading the findings of fact, I can tell what happened. The biggest problem is that often investigators don't tell a narrative story."

For each finding of fact, you must reference an enclosure. But your treatment should be so clear that the reader doesn't need to look up an enclosure to understand what happened.

Guidelines for Findings of Fact

"There are lots of extraneous details—false leads, tangents, dead ends, background information, etc.—that you'll uncover in your investigation, which shouldn't be in the report."

—CHIEF LEGALMAN

- State each fact with definiteness.
- Number each finding of fact and make sure you support each with one or more specified enclosures (one of which might be the observations of the investigating officer).

- Include a specific finding of fact as to the time and cause of any death, supported by a certificate of death or statement by a doctor or medical officer. (Again, don't hold up the investigation to wait for the certificate.)

- If a finding of fact is based upon your personal knowledge as investigating officer, provide the basis for your personal knowledge in a signed memo for the record. Make sure you've questioned all material witnesses. (If not, explain why not in the preliminary statement.)

- Don't include extraneous information. While investigators often err by leaving information out of an investigation, you can also include too much. Officials at one major command recounted a report that was virtually unreadable because it had 2,300 findings of fact—many of them simply unnecessary.

- On the other hand, make sure to record the right kinds of data for the kind of investigation at hand—road and visibility conditions in an automobile accident, alertness of the watch team in a collision, and so forth. Again, checklists for each kind of investigation will help you know exactly what to look for in each case. Double-check these checklists as you draw up your findings of fact.

Example: Poor Findings of Fact

The excerpt below is representative of reports that fail to tell a coherent story. The writer lists facts randomly, showing neither chronological nor logical order, so it's very hard to tell exactly what happened by reading the report. As it turns out, endorsers had to add several additional findings of fact (speed of the vehicles, the speed limit, and others) to this report.

FINDINGS OF FACT

1. That LCDR W— did not hold U.S. Government Motor Vehicle Operator's Identification Card (CT-14) in his possession at the time of the accident. (encls (2), (5))

2. That immediately following the accident, the Fort Polk Military Police were notified. (encl (2))

3. That Mrs. Mary J— was within posted speed limits on Magnolia Avenue. (encl (2))

4. That LCDR W— did fail to yield right-of-way to westbound traffic on Magnolia Avenue after having made a complete stop on General Lee Boulevard. (encls (2), (5))

5. That LCDR W— was on official business at the time of the accident. (encls 121, 151)

6. That LCDR W— holds a valid state driver's license. (encl (2))

7. That LCDR W— observed pavement markings (stop line) on General Lee Boulevard. (encls (2), (5))

8. That LCDR W— made a sworn statement to Fort Polk Military Police concerning the accident. (encls (2), (5))

Example: Good Findings of Fact

The report below follows rough chronological order and tells a coherent story. When it has to shift from the events to describe one of the drivers and a vehicle, it follows a logical train in that discussion, too.

FINDINGS OF FACT

1. That Lance Corporal A— was involved in a motor vehicle accident at 1600 on 22 March 2022. (encl (6))

2. That the accident occurred on the Elysian Expressway, Shreveport, Louisiana. (encl (6))

3. That Lance Corporal A— was driving eastbound on the Elysian Expressway when he lost control of his vehicle. (encl (6))

4. That Lance Corporal A— lost control of his vehicle because of a blowout in a tire on his vehicle. (encls (5), (6), and (8))

5. That the blowout caused the vehicle Lance Corporal A— was driving to cross the cement median and collide head-on with Ms. J—'s vehicle. (encls (6) and (8))

6. That there were only two vehicles in the subject accident. (encl (6))

7. That Lance Corporal A— was driving a 1989 Ford Mustang bearing Louisiana license plate number 579R203. (encl (6))

8. That the vehicle identification number for the vehicle that Lance Corporal A— was driving is 4D05V465698. (encl (6))

9. That there is no record that Lance Corporal A— is the registered owner of the vehicle he was driving. (encl (6))

10. That the Police Officer who investigated the subject accident ran a check on the license plate on Lance Corporal A—'s vehicle to determine if the vehicle was stolen or had any outstanding citation. Officer M— discovered that the license tag on Lance Corporal A—'s vehicle was assigned to a 2003 Volkswagen Golf owned by a Mr. Ralph A. T—. (encl (6))

11. That when questioned by Officer M—, Lance Corporal A— indicated that he had found the license plate and placed it on his vehicle. (encl (6))

12. That Lance Corporal A— was cited for a total of six traffic law violations, consisting of no driver's license on person, no brake tag, no proof of liability insurance, switched license plates, no license plate, and no registration papers. (encl (6))

Especially in longer reports, you can divide the findings of fact into sections that are logically complete in themselves. In a report on damage to an aircraft, damage that occurred while it was being towed about the hangar spaces of an aircraft carrier, for example, the investigating officer divided the findings of fact into these coherent sections:

Environmental Conditions

Personnel Qualifications

Equipment Condition and Documentation

Circumstances Surrounding the Damage

OPINIONS

Opinions are logical inferences that flow from the findings of fact. List only those opinions required by the appointing order—that is, those required by regulations (such as the various "Line-of-Duty and Misconduct" and "Investigation of Specific Types of Incidents" sections of the JAG Manual)—or opinions naturally pertinent to the case.

A good opinions section will seem to flow so naturally from the findings of fact that the opinions seem virtually self-evident.

The biggest mistake made in this section is to begin with preconceived opinions and try to prove them despite evidence to the contrary. Be sure to be open-minded in your investigation, follow where the facts lead, and dig deeply enough to get those key facts that bear significant inferences. Beyond those basics,

- number each opinion separately;

- support each opinion by explicit reference to one or more findings of fact (abbreviated "FF");

- don't confuse facts with opinions (unfortunately very common, with opinions appearing as findings of fact); and

- remember that you must include an opinion on LOD/misconduct in a case involving injury or disease, but that you never include these in an investigation into the death of a service member (see the section on LOD/misconduct above).

Example of Opinions Sections

The following Opinions section is from an LOD/misconduct investigation; note the reference to specific findings of fact in each opinion.

OPINIONS

1. That, due to the length of LCpl K—'s unauthorized absence prior to his injuries (seven days), his absence materially interfered with the performance of his required military duties. (FF (1), (2), (4))

2. That LCpl K—'s injuries were not sustained in the line of duty. (FF (1), (2), (4))

3. That LCpl K— was under the influence of alcohol. (FF (6))

4. That the minor injuries received by LCpl K— during the motor vehicle accident were proximate results of the influence of alcohol and demonstrated a reckless disregard of the consequences. (FF (4), (5), (6))

5. That the minor injuries received by LCpl K— in the motor vehicle accident were due to his own misconduct. (FF (4), (5), (6))

6. That LCpl K— handled a firearm in a grossly negligent manner and demonstrated a reckless disregard of the consequences. (FF (6), (8), (9))

7. That LCpl K— willfully violated a law of the state of North Dakota by assaulting a police officer with a firearm. (FF (9), (15))

8. That the injuries received by LCpl K— from gunshots were due to his own misconduct. (FF (6), (8), (9), (11), (13), (14), (15))

RECOMMENDATIONS

Make recommendations only if the appointing order or the JAG Manual specifically directs you to do so. These recommendations should flow clearly from the expressed opinions and findings of facts, and they may suggest corrective, disciplinary, or administrative action. In addition,

-   make your recommendations as specific as possible;

-   make sure your recommendations are reasonable and just; and

-   see that your recommendations are practicable—that is, that they can be carried out.

Also, realize that your recommendations are not binding on any reviewing authorities; they will undergo a thorough review. As investigating officer, you won't bear the whole weight of the judgment in any particular case. If you recommend punitive charges or letters of reprimand, see the next section on follow-up documents to prepare.

Examples of Recommendations Section

Recommendations Involving Personnel

Below is a straightforward, clear section from an automobile accident report:

RECOMMENDATIONS

1. That Petty Officer M— be the subject of some form of punitive action, either NJP or a Summary Court-Martial.

2. That steps be taken to ensure that Petty Officer M— completes a safe driver's training course.

3. That Petty Officer M— be held responsible for all medical expenses incurred as a result of this accident.

4. That Petty Officer M— be charged lost time for the period of time he was hospitalized at County Hospital.

5. That Petty Officer M— be processed for administrative separation for misconduct, either for civil conviction (MARCORSEPMAN 6210.7), pattern of misconduct (MARCORSEPMAN 6210.3), or both.

Recommendations for Changing Procedures

The recommendations below suggest how to avoid another occurrence of a laundry fire such as has just taken place in the ship's laundry.

RECOMMENDATIONS

1. As stated by the Fire Marshal, it does not appear that lint removal every two hours is sufficient to allow proper circulation of air in the dryers. Recommend that NAVEDTRA 414-01-45-81, Chapter 6 be amended to reflect this change. Until this change is made, recommend lint removal be conducted hourly to preclude any further difficulties.

2. Recommend the ship continue to wash and dry laundry in laundry bags, because washing and drying laundry for several hundred personnel without laundry bags could pose a severe accountability problem. When using the open-mesh laundry bags, recommend the following precautions:
    a. Do not overload the dryers.
    b. Do not place recently dried clothes in nylon bags; avoid concentrating the heat.

3. Recommend that laundry personnel be instructed to keep the dryer thermostats at the recommended level (140–160 degrees). Furthermore, recommend that laundry personnel be required to man the space for a

minimum of four hours after the completion of the drying cycle to ensure detection of any other fires of the sort that occurred.

4. Recommend including all the above recommendations in the laundry training program and in the daily operation of the laundry.

5. Recommend no disciplinary action be taken, as apparently no deliberate actions caused the fire.

DOCUMENTS AND ENCLOSURES

Assembling the report involves putting the various documents you have collected and created as enclosures to the actual report of investigation. As mentioned above, make the written appointing order the first enclosure. Subsequent enclosures should contain all the evidence developed in the investigation as well as charge sheets and punitive letters of reprimand, if recommended.

- Make each statement, document, or exhibit a separate enclosure. Ensure you've completely identified each document.

- Number the pages of a lengthy enclosure to help the reviewer. This way you can specifically reference the relevant passage (by page number) in the finding of fact, as in: "15. The Ferguson vehicle was traveling in excess of 50 mph. Enclosure (10), p. 7; Enclosure (14), p. 3; Enclosure (15), p. 1." You can also tab and highlight pertinent passages in any enclosure that is particularly bulky.

- If your personal observations provide the basis for any findings of fact, attach as an enclosure a memo for the record of those observations.

- For every witness's statement, consider laying a foundation in that statement, either by preface or in the questions asked, to explain to the reviewer why the witness can speak competently on a particular subject.

- Because handwritten documents are often illegible, have witnesses' statements printed or typed. Whenever possible obtain signed, sworn statements. Sworn statements carry greater weight with readers and reviewers than unsworn statements. When you cannot obtain a witness's signature, draft your own summaries of a witness's oral statements. Make sure to sign this summary, certifying it is a valid account or an accurate transcript of the interview, if it is.

- Include as an enclosure the Privacy Act Statement for each witness from whom you obtained personal information by direct inquiry. Attach it to the respective witness's statement.

- Include as an enclosure any prior request (and its approval, too) for exceeding time requirements in conducting the investigation.

- Make sure all copies of the report itself and all enclosures are clear and readable.

Signature and Security Classification

Sign your report.

Omit classified material unless inclusion is essential. Assign the whole report the classification of the highest classified material in it. Staff judge advocates and reviewers will often declassify enclosures and investigations whenever possible; still, some reports will have to remain classified. If yours must be, see that you classify and label the whole report appropriately and ensure that you appropriately classify and label each individual finding of fact, opinion, recommendation, and enclosure. Remember to include the proper downgrading instructions (see OP-NAVINST 5510.1).

Addresses and Copies

- Normally address the report to the convening authority (usually the CO). The convening authority will forward the investigation via the chain of command. For details as to all addressees, see your convening authority and the JAGMAN sections "Action by Convening and Reviewing Authorities" and "Disposition of the Record of Proceedings and Copies."

- Provide an advance copy directly to JAG Office in admiralty cases (Code 11), death cases, or other serious cases so the office will not have to wait for all officers in the chain to act before reviewing the initial findings.

- Provide a copy for each intermediate addressee.

- Make sure that copies to all addressees include all the enclosures. Ensure all photocopies are legible and securely fastened.

- Keep a copy for yourself, unless especially sensitive or classified.

EXAMPLE OF A COMPLETE REPORT OF JAGMAN INVESTIGATION

Figure 9.2 is an example of a complete investigation. This one is relatively brief compared to others. Names and some other details are fictional, but it is based on an actual report.

FIGURE 9.2. *Complete JAGMAN Investigation Report*

23 Feb 2022

From: LT Joseph L. Wilson, USNR, 2305

To: Commanding Officer, USS PIEDMONT (AD 17)

Subj: INVESTIGATION TO INQUIRE INTO THE CIRCUMSTANCES SURROUNDING THE FAINTING OF ET2 SYDNEY LEE SAILOR, USN, ON BOARD USS PIEDMONT (AD 17) DURING GENERAL QUARTERS ON OR ABOUT 1415, 10 FEBRUARY 2022

Ref: (a) JAGMAN
        (b) OPNAVINST 5100.20C

Encl: (1) Appointing Order dated 20 Feb 2022
        (2) CIC Watchbill for 7–27 Feb 2022
        (3) Deck log of USS PIEDMONT (AD 17) 701R09 time 1326 to 1427
        (4) Statement of ET2 Sydney Sailor, USN
        (5) Statement of IT R. B. Rome, USN, Operations Officer, USS PIEDMONT (AD 17) TAO
        (6) Statement of ENS E. F. Snyder , USNR
        (7) Statement of ENS E. C. Johnson, USN, CIC Officer, USS PIEDMONT (AD 17)
        (8) Statement of ET3 V. I. Shirley, USN, JA Phone Talker
        (9) Statement of LTJG R. S. Reynolds, USNR
        (10) Statement of HMC R. C. Jefferson, USN
        (11) Statement of HM2(SW) Y. B. Murfree, USN
        (12) Statement of HM2 C. C. Bruce, USN
        (13) Statement of IT D. T. Daniel, MC, USNR, Medical Officer, USS PIEDMONT (AD 17)
        (14) Medical Department Log Book 0730, 10 February 2022 to 0800, 11 February 2022
        (15) SF600 Chronological Record of Medical Care ET2 Sailor on 10 February 2022
        (16) NAVMED 6500/1 Report of Heat/Cold Casualty ET2 Sailor

PRELIMINARY STATEMENT

1. Following enclosure (1) and in accordance with references (a) and (b), an informal investigation was conducted to inquire into the circumstances surrounding the fainting of ET2 Sydney Sailor, USN, on board USS PIEDMONT (AD 17) during General Quarters on or about 1415, 10 February 2022. All relevant evidence was collected. The investigator met all directives and special requirements set out in enclosure (1).

FINDINGS OF FACT

1. On 10 February 2022 ET2 Sydney Lee Sailor, USN, was on active duty and assigned to the USS PIEDMONT (AD 17). (encl (2))

2. On 10 February 2022 the USS PIEDMONT was steaming from Mayport, Florida, to Norfolk, Virginia. (encl (3))

3. On 10 February 2022 ET2 Sailor was standing underway log watch at her General Quarters station in the Combat Information Center. (encls (2) and (4))

4. Ventilation had been secured because of the drill, and CIC was described as "very hot and stuffy." (encl (4))

5. Material condition "Circle William" was improperly set in that the recirc system R-03-43-2, classified "William," was also secured at the time of the incident. (encl (6))

FIGURE 9.2. *(continued)*

6. ET2 Sailor "became overheated and was perspiring profusely and felt weak and sick. Because of the importance of the drill and because everyone else was uncomfortable [she] was reluctant to take off the MK V gas mask and anti-flash any earlier." (encl (4))

7. ET2 Sailor was instructed by LT Rome to take off her MK V gas mask but became unconscious before being able to do so. (encl (4))

8. ET2 Sailor was caught as she became unconscious and lowered to the deck. Her MK V gas mask was removed. (encls (5) and (7))

9. ET3 Shirley, JA phone talker, relayed "Medical emergency in CIC, not a drill" to D.C. Central. Medical emergency was called away on the 1MC at 1411. (encls (2) and (8))

10. When medical emergency was called away, LTJG Reynolds, HMC Jefferson, and HM2 (SW) Murfree responded from the Forward Decontamination Station with a stretcher team. IT Daniel, Medical Officer, also responded. (encl (9))

11. ET2 Sailor was lying on the deck, alert and conscious. (encls (9), (10), and (11))

12. ET2 Sailor had her feet elevated and was conscious and responsive when the medical officer arrived at the scene. (encl (13))

13. ET2 Sailor was evaluated for injuries and was able to walk from CIC with some assistance by medical personnel. (encls (9), (10), (12), and (13))

14. ET2 Sailor again lost consciousness at the bottom of the first ladder. She was transported to Medical in a Neil-Robertson stretcher. (encls (9), (10), (11), and (12))

15. ET2 Sailor arrived at sick bay in a Neil-Robertson stretcher at 1415. (encls (14) and (15))

16. ET2 Sailor's vital signs were normal. (encl (15))

17. ET2 Sailor was diagnosed and treated for heat exhaustion, i.e., VASOVAGAL SYNCOPE SECONDARY TO HEAT STRESS. (encl (15))

18. Medical Officer's notes state: "Past medical history unremarkable except that patient states that she and other members of her family have a tendency to 'pass out easily.'" (encls (15) and (16))

19. ET2 Sailor had eaten lunch before General Quarters, had had some 9 hours of sleep in the past 24 hours, and had drunk 4 to 5 cups of coffee and 6 glasses of water in the 12 hours prior to her illness. (encl (16))

OPINIONS

1. That ET2 Sailor became unconscious as a result of mild heat exhaustion. (FF (17))

2. That ET2 Sailor was motivated to participate in the drill and would not have fainted if she had spoken up sooner. (FF (6))

3. That the combination of material condition and battle dress probably precipitated the illness, but that ET2 Sailor is prone to fainting and might have fainted as a result of battle dress alone. (FF (4), (5), and (18))

4. That the illness occurred in the line of duty and not as a result of her own misconduct. (FF (3), (6), and (7))

5. That there is no likelihood of permanent or recurring illness or disability as a result of the single episode, and that claims against the government are not warranted. (FF (16–19))

FIGURE 9.2. *(continued)*

6. That first aid rendered by personnel at the scene was correct, that the medical emergency was promptly and correctly called away, and that medical department response was prompt and correct. (FF (9–15))

RECOMMENDATIONS

1. That ET2 Sailor be counseled not to tax herself to the point of illness during a drill scenario.

2. That supervisory personnel in CIC and the Communications Center ensure that recirc system R-03-43-2 remain energized so long as electrical power is available to the controller.

JOSEPH L. WILSON

INVESTIGATION ENDORSEMENTS

The chain of command will review your report thoroughly and repeatedly. That's why you have to be sure that you've done it all as well as you can.

The purpose of an endorsement is to give the reviewer's point of view on the matter under investigation and to make sure that you've followed all technical procedures (for example, that you've observed all the rights and given the appropriate warnings). In other words, an endorsement is both a check or review of the report and an opportunity for the reviewing authority (the first of which is usually the CO) to make comments.

The reviewing authority will comment on the soundness of the findings, opinions, and recommendations and will also advise what follow-on actions have been taken in relation to the case. A CO will typically state which recommendations he or she has approved and acted on, what further steps beyond the recommendations have been taken, what disciplinary action the unit has initiated, and what administrative improvements the command has decided on.

Figure 9.3 is a fictionalized version of an actual endorsement. Notice the use of plain language principles such as active voice, making it read more clearly than many endorsements often do.

Each authority to whom you route a JAGMAN investigation forwards it by endorsement. If an authority decides the investigation has major errors, that person can turn the report back down the chain for further inquiry or for corrections. The endorser can also simply correct minor errors, add any additional necessary findings of fact or opinions, approve or disapprove of the conclusions and recommendations (perhaps adding others), and send the package on.

This process doesn't mean you're completely off the hook for any errors you may have made in the investigation. After forwarding the report, the endorser will send a copy of the endorsement back down the chain to your CO's reporting senior(s)

FIGURE 9.3. *Endorsement to a JAGMAN Investigation Report*

5800
Ser 12/0037
15 Sep 2021

FIRST ENDORSEMENT on LT John L. Wilson, USNR, 1310, ltr dtd 11 SEP 21

From: Commanding Officer, USS GABRIELLE GIFFORDS (LCS 10)

To: Commanding Officer, Naval Legal Service Office (Claims Department), U.S. Naval Station, San Diego, California

Subj: INVESTIGATION TO INQUIRE INTO THE CIRCUMSTANCES SURROUNDING OVERSPRAYING OF VEHICLES IN THE VICINITY OF USS GABRIELLE GIFFORDS DURING THE PERIOD OF 10 AUGUST THROUGH 15 AUGUST 2021

1. Readdressed and forwarded.

2. The Deck Department, USS GABRIELLE GIFFORDS, and Port Services, U.S. Naval Station San Diego, did take reasonable precautions to place drivers in the vicinity of the ship on notice of spray-painting operations. Although most claimants knew or should have known about the potential for damage to their vehicles, it is apparent that some vehicles incurred damage after reasonable preventative steps had been taken.

3. Spray painting is a fact of life for Navy ships. This ship must be painted regularly, and the methods of accomplishing this evolution during the month of August were appropriate. In addition, significant efforts were taken to provide information to those who park and work in the vicinity of USS GABRIELLE GIFFORDS.

4. During the period in question, the ship was fully engaged in Selected Restricted Availability, which required extensive refurbishing, repairs, and painting. The Availability was the primary mission of all personnel assigned to USS GABRIELLE GIFFORDS at the time.

5. Typical paint overspray consists of a light mist that is normally removed from a vehicle with a rubbing compound and an hour or two of buffing. It was this method that I used successfully on my own black sports car to remove overspray from the same painting operations.

6. Subject to the foregoing, the findings of fact, opinions, and recommendations of the investigating officer are approved.

George Z. Watson

Copy to:
OJAG (Code 11, Admiralty)
Naval Station San Diego
LT John L. Wilson

---

first. On a staff, an endorsement will usually go on the read board, so that many eyes will see seniors' criticisms of the report you've written. The endorsement will eventually get back to your own CO and finally to you. You can be sure that your boss will notice (and certainly not appreciate) any mistakes you've made.

As a cautionary example, several years ago a loss of $150 was discovered aboard a destroyer, an event that caused the relief of the ship's store operator (a petty officer third class) and some considerable embarrassment to the supply officer. But that's not where the embarrassment ended. The third endorsement to this investigation, signed by the commander of a cruiser-destroyer group, had this to say:

> The overall lack of quality of the basic investigation, particularly the absence of written statements from the principals, is noted with concern. . . . Commanding Officer, USS VESSEL (DDG XX) is directed to ensure compliance with reference (a) in future investigations.

This critique was sent on to the JAG via two other major naval organizations (standard practice at the time), with copies to the ship's DESRON commander, the ship's CO, and the lieutenant who wrote and signed the original investigation.

As the O-6 JAG at a major type command commented, too often an investigating officer will call up the staff JAG when first seeing such an endorsement and ask: "Why didn't you call me? Why did you have to put the gig in print? My skipper is going to hit the overhead when he sees this!" You have to remember that once it's in print, it's official, and that the staff JAG works for the O-6 or flag, not for you. (The process is slightly different in the Marine Corps. Marine legal officers commented that they make a policy of doing the same review for the battalion officer as they do for the general, thus solving perceived problems before an investigation reaches the general's desk.)

What will the endorser look at? Anything and everything suggested above and perhaps other details, too. Before submitting your report, check it over one last time to make sure you've met all the requirements. If you can say yes to all the questions in the checklist in figure 9.4, you can be pretty sure you have a decent report.

FINAL CHECKLIST ON JAGMAN INVESTIGATIONS

> *On two different audiences for a JAGMAN investigation: "There are two dragon's mouths to avoid, and they read the investigation in different ways. The staff JAG, the attorney, looks to see if you followed the regulations, if you jumped through all the hoops. Was the checklist completed, was the specific language used? The flag, on the other hand, looks for readability and the bottom line. And he doesn't appreciate legalese."*
>
> —MARINE JAG MAJOR

As you prepare to submit your final report of investigation, use the checklist in figure 9.4 to review the document and make revisions.

FIGURE 9.4. *JAGMAN Investigation Report Final Checklist*

✓ Does your report "answer the mail"? Have you carried out what the appointing order directed you to do?

✓ Do all the parts mesh? Does the preliminary statement properly introduce the rest of the report? Do the findings describe the basic facts? Do the opinions and recommendations logically follow through?

✓ Does every fact have an enclosure and every opinion a reference? Are the opinions and recommendations reasonable?

✓ Has the investigation been as thorough as you could reasonably expect?

✓ Is the report technically complete and correct in all its details? Have you classified it correctly? Have you identified all the witnesses?

✓ Are all the enclosures in place, properly marked, and highlighted?

✓ Are all the copies (including reproduced copies of enclosures) readable? Have you made enough copies of the report?

✓ Have you shown this report to someone else (a knowledgeable reader and good critic)? Can that person follow it all? Does it make sense to this reviewer?

✓ Does the report needlessly present a poor image of your command or any individuals involved in the investigation?

✓ If you were the officer charged with ensuring that the government's interest had been looked after, would you think all required duties and responsibilities had been carried out?

✓ On the other hand, if you were the person whose acts were under investigation, would you think you had gotten a complete hearing and a fair shake?

10

Social Media
and News Writing

> *"Contributions with the pen have been significant for even the busiest military*
> *professionals. The written word represents another 'weapon' for use in achieving*
> *professional objectives and making an impact on society at large. For patriots,*
> *it remains a valuable tool, readily accessible for use and powerful in effect."*
> —WILLIAM K. RILEY, EDITOR, ARMED FORCES STAFF COLLEGE

EXECUTIVE SUMMARY

Social media and news media writing are inherently public. Writers for the command or simply posting to their own accounts have an opportunity to promote the Navy or Marine Corps' interests, mission, and reputation. They also need to be cautious not to undermine those very things with inappropriate posts or violations of operational security (OPSEC). Taking the distinct audience of the different social platforms into account is an important element in making your posts successful. Commands aiming to establish a social media presence should carefully consider their aims and the way this engagement supports their overall mission. Successful newswriting must also target its audience and follow a structure and style appropriate to readers. In addition, the writing style for internal Navy correspondence and other documents differs from the standard journalism style that writers for news media follow.

Writing for social or news media is distinctly different from most of the other composition discussed in this book. Both types of media play a major role in helping families and the general public feel connected with what Sailors and Marines are doing at sea, out in the field, or in a staff or other administrative role. Both the Navy and the Marine Corps have published a handbook on using social media, each of which is available online and referenced at the end of this chapter and in appendix A.

Writing for Social Media

"Social media writing offers one of the most powerful tools ever seen for communicating, connecting, and shaping public opinion. But with that great power comes great responsibility—and the potential for great blunders and even harm."

—LIEUTENANT COMMANDER,
FORMER NAVAL ACADEMY ENGLISH INSTRUCTOR

AUDIENCE, PURPOSE, CONTEXT

Social media is fundamentally public. Even what you share in a private chat or group can easily be passed along to many others. This is true for email as well, of course, but the wide, public dissemination of what others share is—at least for many people—practically the essence and purpose of social media.

Audience

With social media, as with every other type of writing discussed in this guide, you need to consider who your intended audience is, what purpose you want to accomplish with the post, and what aspects of the broader national context will influence (for better or worse) how your words and images are interpreted and understood. Most of your naval writing (such as evaluations, messages, and official email) targets audiences within the Navy. Social media writing includes them, but it also includes the wider public. Even a Facebook post highlighting some Sailors promoted from petty officer second class to first class, though primarily intended for members of the command and their immediate families, needs to be written with extended family members, friends, and the wider public to whom the post may be forwarded in mind.

Purpose

Simply sharing the information of the specific event is not the full purpose of a posting. Beyond this, at least with official social media writing for the Navy, it is to cultivate a positive relationship between the service and the public, including the families of the women and men assigned to a command. More broadly, this relationship includes the public image of the Navy. Whether drafting a tweet or a Facebook or Instagram post, choose your content and style carefully with this general objective in mind.

Context

For social media writing and newswriting (discussed later) the larger political, economic, and cultural context that is on the minds of your readers will influence the way your posts are interpreted and understood, even if you think they have nothing to do with politics, economics, or culture. Before you post something, take a moment to imagine different ways these broader elements could distort your intended message as written in the readers' minds. In addition, smaller localized aspects of context can also affect interpretation: time of day or time of week, payday, a holiday, the ship's schedule, and other details affecting how your Sailors and families are feeling. Recognizing these things and crafting your posts to capture readers' attention amid competing posts and concerns is vital to their success.

If you are a CO or commander of a unit, as Navy guidance states, you should think about what you want to accomplish before you launch a social media presence for your command. Consider what communications will help your command achieve its mission. Who are the most important audiences for a given set of information, and which platform will best reach them?

WRITING FOR DIFFERENT PLATFORMS

"Do you want to communicate with your Sailors, Navy civilians, command leadership, family members, the local community, a broader DoD audience, the American public, or another group altogether? Do you have the content and personnel—both now and long term—to routinely engage with those audiences?"
—*U.S. NAVY SOCIAL MEDIA HANDBOOK*

If you already use multiple social media platforms, you will know that each one has a distinct flavor. The general user demographics, the types of posts, and the tone of the writing differ somewhat based not on rules but on trends and expectations that have developed over time for each of them. If you are managing a social media presence for your command, you will likely need to adapt the same information for posting on multiple platforms. The three primary platforms the Navy is using for official social media as of this writing are Facebook, X (formerly Twitter), and Instagram.

Whichever platform you are using, and whether you are writing for it in an official capacity or in your personal account, you have a professional responsibility to represent the Navy or Marine Corps well. One of the most challenging aspects— read greatest potential pitfalls—of being on social media lies in responding or reacting to comments. A person may comment in an ugly way on a legitimate and well-written post. Keep your emotions in check and be cautious in your response. In official posts representing your command, especially for any issue with high visibility, use a pre-prepared ("boilerplate") response approved by the legal office in advance. Also, do not say anything on social media that you are not 100 percent sure of.

Facebook

Facebook is, of course, the most broad and widely used platform. But it has become less popular with younger people. As of early 2021, 69 percent of adults said they used Facebook, the most of any social media platform other than YouTube. Across the board, more women than men (83 percent vs. 77 percent) reported using Facebook, according to Pew Research. Its largest user age group was adults twenty-five to thirty-four years old, with about 60 percent of them being male. Adults from eighteen to twenty-four and thirty-five to forty-four years of age made up the next-largest portion of Facebook users, also with slightly more men than women, according to the website Statista.

X (formerly Twitter)

The most distinctive feature of Twitter, rebranded in 2023 as X, is the brevity of posts: 280 characters as of early 2021. Yet very few posts are actually longer than the old limit of 140 characters. Posts (formerly called "tweets") must be not only brief—if you want them to succeed—but also to the point. With this tight limit on length, if you want your message to catch readers' attention (amid myriad others they can read), you must choose each word carefully. For detailed advice on this process, see chapter 1 on revising your sentences for concision.

Just as Facebook and Instagram thrive on users' tagging others and sharing posts, X thrives on people reposting what others post. Unlike a brief email that you know people *might* forward, however, you should *expect* your messages to be reposted, and you should craft them with this goal in mind. Consider how something sounds on a first read, then consider how it sounds on a tenth read. Try to imagine how readers with different opinions and biases, diverse backgrounds and levels of understanding, might receive it. Not that you can make them all happy, but in an official post for your command, you should aim to promote the reputation and best interests of the Navy or Marine Corps and their Sailors and Marines as effectively as possible.

Another important aspect of organizational X accounts is that, largely because of companies that use the platform for marketing, such users often create a "persona," a style and tone of writing tied to the mission. Although a Navy command does not have to be ultraserious in every post, consider where on the spectrum of serious to frivolous and witty you want this organizational persona to fall.

Lighthearted can help engage families and the general public, and you can still maintain decorum and practice OPSEC while sounding human.

Instagram

Instagram's largest demographic comprises eighteen- to twenty-nine-year-olds, and its smallest audience is adults over sixty-five. Its users also tend to expect more lighthearted posts, so if you put up something entirely serious, it could be misinterpreted. Of course, Instagram is far more visual than verbal. But you can still put the written word to good use. A carefully worded caption to your photo or "story" (as Instagram video posts are called) can direct the audience's understanding and interpretation of the visual element. As of 2023, the character limit for captions on Instagram is 2,200 characters. Although that figure allows you to write quite a lot, the attention span of your audience is limited, especially as users skim through their feed. Keep captions brief. If you want to reach your Instagram audience with positive visual posts about the command, don't push a lot of text. Write just enough to make the photo or video clear in the way you intend.

NAVY AND MARINE CORPS REGULATIONS ABOUT USING SOCIAL MEDIA

If you're expressing a personal opinion of any kind, it's your responsibility to make clear that you are not speaking for the Navy and that the stance is your own and not representative of the views of the Navy. Furthermore, the same relationship and fraternization boundaries apply on social media as apply in the physical workspace. Navy and Marine Corps leaders must be especially careful not to cross those boundaries when posting or when responding to posts of those in their chain of command.

Another crucial element of Navy social media policy lies in the fine line between the personal and the professional when it comes to what you post online. Keep the following guidance in mind.

No Sailor should communicate on social media or elsewhere in a way that may negatively affect herself or himself or the Navy. It's often hard to distinguish between the personal and the professional on the Internet, so Sailors should assume any content they post may affect their personal careers and the reputation of the Navy more broadly. They should not engage in any conversations or activities that may threaten the Navy's core values or operational readiness.

Content that is defamatory, threatening, harassing, or discriminatory on the basis of race, color, sex, gender, age, religion, national origin, sexual orientation, or any other protected status is punishable and must be avoided. The Internet doesn't forget; online habits leave digital footprints. Take caution when posting content, even if you think you're doing so in a private, closed community.

This caution includes so-called private or closed groups as well as public ones. Understand that *everything* you post is recorded, and for purposes of OPSEC and other potentially harmful or inappropriate posts, you should be just as cautious in private groups as in open ones.

WRITING FOR THE NEWS MEDIA

"Writing is the way to reach those people when they are not at hand and you have no way of getting to them. Talking is like a 24-pdr muzzle loader: terrifically effective at very short range. Writing is like an airplane or a missile: It permits you to be effective over the horizons of space and of time."

—FRANK UHLIG, FORMER PUBLISHER,
NAVAL WAR COLLEGE REVIEW

Who besides official mass communication specialists (MC) and public affairs officers (PAO) need to know about newswriting? Quite a number of people should possess this skill, as it turns out. As with the use of social media to communicate directly with families and the public, COs want to tell the good story of their commands and get their people recognition. COs, XOs, and CDOs may have to approve and release news stories; Marine Corps unit information officers have to put out information on their units; and many Navy officers on ships, air squadrons, or other stations are usually assigned as collateral-duty PAOs.

Strong news releases can lift crew morale. One educational services officer who also did PAO work on an amphibious assault ship began putting an article in the base newspaper every month—the CO loved it. When the ship was returning from a deployment, 300 award recipients spelled out the vessel's hull number on the flight deck for a photo, which appeared in the Norfolk and Little Creek base newspapers—the awardees loved it. The CO went into the engine rooms during an inspection and got another article with photos into the base paper—and the engineers (who never had such recognition) loved it, too.

Besides enhancing crew morale and pleasing the CO, news or feature writing can do a real service in informing the public, not only the families and friends served by base newspapers but also the wider naval audience reached by such papers and magazines as *Navy Times, All Hands, Naval Aviation News, Navy Reservist*, and *Leatherneck*. Of course, there are also many civilian outlets for naval news, from the local paper and other area media on up. Articles that reach the readers of these publications can greatly help—or hurt—the Navy's or Marine Corps' reputation.

Newswriting, like that for social media, is distinctly different from most of the other compositions covered in this book. It requires a "wholly different style," according to the CO of one training command. Part of the difference lies in the nature of the subject matter. While there are different kinds of newswriting—and some elements are more important to one kind than another—any genuine news item should contain all or most of these basic ingredients:

- Something must HAPPEN!
- It must be timely.
- It must be significant.
- It must have local interest.

The item will be even more valuable as news if it contains one or more of these elements as well:

- Humor
- Conflict
- Human interest
- Well-known personalities
- Suspense

Recognizing news is one thing; finding it is another. Some stories will come to a writer as gifts. But you will obtain most of your significant news only by lots of legwork, a little ingenuity, some imagination, and an organized routine. Don't wait for news to find you—go seek it out.

Military writers make use of two standard kinds of journalistic writing: news releases and feature stories. These differ primarily in their timeliness and relationship to the reader's knowledge and interest.

The news release and feature require sharply different techniques. Both are useful for naval writers to learn, not only for use in newswriting but also for the spin-offs they have in day-to-day staff and organizational writing. We'll look at each of them in some detail.

"I keep six honest serving-men / (They taught me all I knew); / Their names are

What and Why and When / And How and Where and Who."

—RUDYARD KIPLING

THE NEWS RELEASE

Structurally, the news release follows the order of decreasing interest, often called an inverted pyramid, giving the story's most important idea at the beginning and then gradually narrowing with specific details. With this approach, you assume a reader might stop reading anywhere in the story—or the editor may need to shorten the article by chopping off part of the end. You want to get the most important information to the reader at the beginning. A writer begins the story with a basic summary of the entire article and the most crucial facts, then elaborates on those facts throughout the article.

News releases have three parts: the lead, the bridge, and the body. The most important of these is the lead, often spelled "lede" by journalists.

FIGURE 10.1. *Inverted Pyramid Model for Structuring a News Story*

(order of decreasing interest)

LEAD
BRIDGE
BODY

Opening with the Lead

Although reader interest is always a consideration, you don't design a lead primarily to get someone to read a story. Headlines have more of that function, and the subject matter itself is probably the most important factor in determining which news stories any particular person reads. Rather than being an enticement, a lead is better defined as an effective summary that conveys all the basic facts at once, allowing the reader to scan the paper and decide which stories to read through for more details.

The lead is usually a single sentence of some twenty to thirty words and can be of several types, each of which has special usefulness. You categorize a lead by which of the five Ws and one H it emphasizes. (Although more than one of these should appear in it, just one receives special emphasis). Is it a *who, what, where, when, why,* or *how* lead?

USS Roosevelt (DDG 80) Begins Second FDNF-E Patrol

The *Arleigh Burke*–class guided-missile destroyer USS *Roosevelt* (DDG 80) departed Naval Station Rota, Spain, to begin its second Forward-Deployed Naval Forces–Europe (FDNF-E) patrol, March 29, 2021.

— MC2 Rumple, 29 March 2021

Suicide and COVID-19: How Navy Region Southeast Is Fighting Back

Navy Region Southeast worked with LivingWorks Start to offer a free online suicide awareness course to military personnel, family members, and DoD

civilian employees. The free 90-minute online program provides person-
alized information about suicide prevention and the meaningful actions
someone can take to help keep others safe.

— Navy Region Southeast Public Affairs, 10 March 2021

Navy Approves Alternate Rank Tabs for Type III Work Uniform

The Navy has authorized Sailors the option to wear the black Cold Weather
Parka (CWP) sleeve-style rank insignia with the Navy Working Uniform
Type III (NWU Type III) in non-tactical environments.

— MC1 Mark D. Faram, Chief of Naval Personnel Public Affairs, 1 November 2020

A good lead, then, offers the basic information of the story in quick summary
form—like this:

Eight Sailors on two fifty-foot Navy shuttle boats rescued two people from
the Mississippi River at 6 A.M. this morning.

— NSA New Orleans Release 95-0838

This lead above is a what lead. Writing leads like this one requires practice.

You should write or at least plan out most of your story before writing the lead so
you know what your article will say. Before writing a lead, list the who, what, where,
when, why, and how of your story. Then consider the story from your audience's
viewpoint. Which emphasis will likely most resonate with your reader? Which
might be most interesting, or hardest hitting? Theoretically, by emphasizing any
one of these aspects, you could compose six different summary leads. Yet when and
where—important as they are to a news story—don't usually provide the best leads.
On the other hand, there are two different kinds of who leads. Altogether, you have
five strong possible leads for a story:

1. A *what* lead—usually the most important element in hard news stories:

   The destroyer USS *Deyo* (DD 989) arrived in its new home port of Norfolk
   May 22, relocated from Charleston, S.C., as a result of the base closure and
   realignment decisions announced in 1993.

2. A *who* lead—especially useful if the who is well known or holds an important position:

   The Assistant Secretary of Defense for Reserve Affairs, Deborah R. Lee,
   will meet with senior leaders of the Marine Corps Reserve in New Orleans
   Friday at 2:00 P.M.

3. An *impersonal who* lead—used if the who is not well known or not important; the person is identified in the bridge or the body of the story:

> Navy wrestlers won 4 of the 17 medals taken by U.S. wrestlers at the Conseil International du Sport Militaire in December.

4. A *why* lead—why is sometimes the most interesting element:

> Because of a shortage of instructors in the Naval Dental School, the student-instructor ratio will be increased to 16:1 next month.

5. A *how* lead—occasionally the how is the strongest possible lead:

> By dropping a bomb down the chimney, an escaped arsonist blew up the town judge's home yesterday.

Remember: don't get bogged down in secondary details or get tangled in the chronology of events in a lead; let those elements follow in the body.

The Bridge as the Link

The bridge is a sentence or two that links the summary information in the lead to the detailed information of the body. Although not always required, it helps a writer avoid cluttering the lead with secondary facts. The bridge serves the lead because the kind of lead usually determines the type of bridge that follows.

For example, a bridge following an impersonal-who lead provides complete identification of the individuals mentioned in the lead.

Lead: An off-duty Marine gate guard stationed here saved a 10-year-old boy from drowning just off the seawall near the base gym.

Bridge: By jumping off the seawall and swimming 50 yards, Corporal Sam Jones saved the life of the unidentified youngster, who had fallen off a sailboat.

Another common purpose is to tell the source of information given in the lead and to supply additional information.

Lead: Vandals in the students' barracks destroyed nearly $12,000 worth of fire protection equipment in the last six months, leaving their fellow students at the risk of injury or death.

Bridge: According to Base Fire Chief Charles W Smith, vandals have torn down smoke detectors, bells, and pull stations that contractors installed in the barracks about six months ago. In addition, 211 fire extinguishers had to be replaced, the chief said.

A bridge following a lead that omits some of the five Ws simply adds other information.

Lead: The harbor tug *Wobegon* (YTB 472) rescued two men from a disabled motorboat that was drifting to sea on the ebb tide early Saturday morning.

Bridge: The motorboat had been without power for three hours when the *Wobegon* appeared about 4:30 A.M. a half-mile southeast of Fort Sumter, Charleston, S.C. The two men in the boat, who were not identified, got the tug's attention by a flashlight SOS.

Other bridges tie a lead back to earlier stories on the same subject or simply add additional facts.

The Body Provides Details in Decreasing Importance

In the body, elaborate on the elements given in the lead and bridge by adding details, in-depth discussion, chronology of events, or quotes on what occurred. In essence, the body retells in detail and descending order of importance the summary facts given in the lead.

Remember these important principles when writing the body of a news release. First, an editor with limited space who wants to use your story will cut from the bottom. Make sure no essential material is left to the end and place the most important information in paragraphs immediately succeeding the lead and bridge. Again, write in descending order of importance, with minor details left for the end.

Second, remember that newspaper columns are very narrow. Dense blocks of type inhibit reading wherever encountered, but narrow columns of news type can make reading even more difficult than usual. Limit your paragraphs to two or three sentences in length.

The Heading

Once you've composed a news release complete with lead, bridge, and body, don't overlook one other part—a news release heading. A good heading can mean the difference between good material being used, lost, or thrown out. Include in this heading these five basic items:

1. The name of the command, which identifies the releasing authority.

2. The contact individual within the command. Provide your name, phone, and fax numbers are very important—they tell the editor how to reach you to clarify information, to confirm some aspect of the release, or to answer other questions.

3. The telephone and fax numbers and email of the command or contact individual.

4. A "slug," or title, to give the editor an idea of the story content.

5. The release date and time. Note the following common terms:
   - FOR IMMEDIATE RELEASE—for hot items
   - FOR GENERAL RELEASE—for a feature and other items with no time element
   - HOLD FOR RELEASE UNTIL—for items mailed in advance of the time/date when the public can be informed
   - DO NOT USE AFTER—for items not accurate or pertinent after a certain date (such as publicizing an open house, for example)

Don't include a cover letter for your release—that's too much official clout, and editors don't have time to read a cover letter anyway. The paper will recognize you're submitting a news release by reading the heading and the title "News Release."

FEATURE STORY

Many naval stories—especially those written in weekly base papers—are feature stories. Commercial newspapers usually cover hard news, while base newspapers, which have limited staff resources, simply can't be as up-to-date as other news sources. Timeliness is still important in some instances. A story about a ship returning from deployment should come out in the base paper within a week of the event; the same applies for changes of command and so on. But in most cases the interest in such stories is no longer on the event itself but on some special aspect of it. So, instead of a "hard news" release, you would write a "feature story" and send it either to a base newspaper or, if it is good enough, to a journal like *All Hands*, *Surface Warfare*, or *Marines*. These magazines aren't interested in "hard news" but rather in human interest, humor, technical information, and military or institutional facts of life. Feature stories on Navy and Marine Corps subjects can help humanize the services for nonnaval audiences. This type of reporting helps citizens relate to and understand men and women in the service. On the other hand, feature stories that specifically target service members and their families—most of the features in base newspapers—help us relate to and understand one another. They deal with our tasks, jobs, environments, desires, loves, challenges, and predicaments. Moreover, by giving fitting recognition to the dedication and deeds of service members and their families, news features help boost individual and unit morale.

The task of the feature writer is first to find that special story worth telling, then to tell it in a way that will prick interest. As a result, the whole approach of a feature story differs from that of the news release.

Writing Feature Leads

The feature lead serves a different purpose than the news lead. Unlike the news lead, it does not deliver all the basic facts at once—instead, it attempts above all to get readers interested, to grip them, and to make them want to learn more. Many different leads can introduce feature stories—there is no standard formula based upon the five Ws. Nor is there a necessary "lead-bridge-body" structure, but instead uses a lead to intrigue, followed by secondary interest, often leading to a story and maximum reader interest. Maybe such a formula is impractical in all cases, but still, material of sustained interest should follow the lead, and a well-crafted ending should wrap up the story.

The following paragraphs illustrate several popular kinds of feature leads connected with several different kinds of stories. These examples are just a few of many possibilities.

The Summary lead opens with facts. It resembles a news lead but is more interesting.

> The path to earning a place as a musician in the Navy or the Marine Corps is far from easy, requiring years of formal training and a rigorous audition gauntlet that ensures that only the very top people who apply are accepted.

The Narrative or Descriptive lead sets the mood, stirs emotions, and gets a reader into the story. The following lead is adapted from a story written by Sgt. David J. Ferrier.

> Not able to sleep, Lance Corporal Zachary Mayo put on his blue coveralls, green T-shirt, and boots and walked out on the catwalk—a place he often went to get fresh air. Mayo remembered later that he had forgotten to firmly shut the ship's heavy, steel, watertight door. As the aircraft carrier USS *America* patrolled the North Arabian Sea, she turned to port. Swinging open, the hatch slammed into Mayo's back, knocking him through the safety rails, and he plummeted nearly six stories into the ocean.
>
> The Osburn, Idaho, Marine survived nearly 36 hours alone and adrift on the high seas.

The Teaser lead captures one's interest by promising something interesting without telling the reader what the story is about. The reader learns more of the subject as the story continues. The following comes from a story about the Navy's first aircraft, the A-1, but doesn't name it until later.

It was a light, delicate aircraft, but in its short life it survived several crashes. And like most pioneers, it had its share of failures and accomplishments.

The Quotation lead is suitable for historical stories or recent events. Use it rarely, then only when dynamic and short quotes are available.

"You really don't know what freedom is until you have had to escape from Communist captivity," said Navy Lt. Deiter A. Dengler, an escapee from a Viet Cong prison camp.

The Direct Address lead states or implies the word "you" somewhere in the first paragraph.

If you need special uniform items from the NEX before the next inspection, don't wait until just a few days beforehand; the new delivery schedule for items at the uniform shop may require over a week's notice.

Structuring the Feature Body

As with the news release lead, authorities often advise that you write the feature lead first before writing a story. The lead will often set the stage for the rest of the article, suggesting a logical structure.

How you go about writing the rest of the story varies. "The bottom line in feature writing," said JOC Jon Cabot in the Spring 1983 issue of *Direction*, "is adding personality or character to any given subject. The easiest way to take a topic and make it a feature is to think in terms of putting the topic on a stage, much like an actor. You then move the topic across the stage through the effective use of quotes and anecdotes until you come to the end of the stage or your summary." Over forty years later, Cabot's insight still provides excellent advice.

Colorful writing and freedom of composition are not only permissible but encouraged in feature writing. This subjectivity differs from writing news releases, in which journalists do their best to be (and to appear to be) objective. "Straight news" writers see to it that no opinions remain unattributed, and they scrutinize their adjectives to ensure no bias has crept in. They do their best to disappear from the story.

In contrast, writers of feature stories are allowed—even encouraged—to state opinions (they usually receive bylines, not only to give them credit for their stories but also to identify the source of whatever opinions appear). They are also urged to write colorfully to keep up the reader's interest. Writers can use several techniques— looking at events from differing points of view, making intriguing comparisons and contrasts, using striking quotations—the possibilities are endless. The best way to learn to write features is probably to read several, studying the techniques in those stories that especially interest you.

The end of a good feature shows as much craft as the beginning. One good way to end is with a choice quotation from an interview; another favorite method is to summarize the key points of the story and comment on their importance. A good feature story keeps the interest of the reader to the very end and, if possible, surprises or delights the reader even in the conclusion itself.

Most official Navy and Marine Corps magazines are written by staff members. Yet this does not seem to be the case with *Approach*, the Navy and Marine Corps "aviation safety" magazine. Its articles are solicited from fliers and maintenance people in both services who have usually had some kind of close call. Because of the frank acknowledgment of difficulties, the typical lively writing style (no doubt enhanced by the editorial staff), and fleetwide interest in naval aviation, the magazine's articles perhaps have more inherent interest than those of any other official publication. Certainly, the "lessons learned" found in *Approach* are quite effective.

The short article in figure 10.2 was written by a Marine Corps captain and was accompanied by a photograph in which he displayed his left hand with its missing finger. Given the arresting image, one almost had to read the article. Note the effective use of four short imperative sentences at the item's very end.

FIGURE 10.2. *Short News Article*

DANGER LURKS

The morning brought with it my annual egress drill from the AV-Harrier. I was wearing all survival gear required for flight, as briefed by the safety officer. Upon strapping into the aircraft, I realized I wasn't wearing gloves. Further, I was wearing my wedding ring. The egress was going fine until I went to release my grip from the aircraft's canopy rail to get down. My ring caught on the large, rearward facing canopy hook, and held fast as the rest of my body continued descending. I felt a jerk, heard a ripping sound, and looked down to discover my ring finger was totally severed and hanging by a thread perpendicular to its normal position.

It looked as if a sock had been rolled off the finger and left only a bloody bone remaining; my top knuckle was completely torn off.

The emergency room doctors and hand specialists determined that all tendons, nerves and blood vessels of the finger had been destroyed and were beyond repair. A few hours after the accident my finger was amputated a half-inch above its base.

An important and costly lesson learned is that no task in naval aviation is routine. A simple egress drill for me turned into a finger amputation. No matter your experience level, there is always an unforeseen danger lurking in the shadows to take advantage of the unsuspecting aviator or maintenance person. Fight complacency. Wear the required safety gear. Remove all jewelry before work. Spread the word.

— by Capt. Matt Vogt, USMC (at the time he was flying with VMA-542), *Approach*, May–June 2006, 11.

Figure 10.3 is an example of a feature story written years ago by a collateral-duty PAO aboard a destroyer. Published in *Soundings*, the base newspaper in Norfolk, it is a good example of a standard feature that a part-time PAO would write. After the heading, the story begins with a quotation lead and then proceeds to tell its story, keeping up interest with details about the Cape Verde Islands and an adept use of several quotations, including a good one to end the story.

FIGURE 10.3. *Feature News Article*

USS COMTE DE GRASSE (DD 974)
Lt. Joe DiRenzo III, Public Affairs Officer Phone 804-444-7552

"Comte de Grasse Makes Port Visit to the Cape Verde Islands"

FOR GENERAL RELEASE

"It is a distinguished pleasure and privilege to host the USS Comte de Grasse (DD 974) here in the islands," commented U.S. Ambassador Vernon D. Penner, Jr., during a luncheon on board the ship as it rested at anchor in the Cape Verde Islands. "We have had other ships, but nothing like this magnificent vessel," continued Penner.

The USS Comte de Grasse, having just left the Mediterranean Sea and duties with the U.S. Sixth Fleet, had traveled more than 1,500 miles to this tiny group of islands 385 miles west of Senegal.

Under the command of Cmdr. Russell J. Lindstedt II, the officers and crew were treated to a port visit vastly different from any they had seen during their five and one-half months in the Mediterranean.

With several small fishing boats on either side, the Comte de Grasse anchored off the port city of Mindelo. "Mindelo's ties with the U.S. are rich ones," commented the ambassador.

"Mindelo is the sister city of Bedford, Massachusetts, and has many exchanges with Bedford," continued Ambassador Penner. "Overall, our ties to this country extend over 900 years and have been strengthened by visits like this." Many Cape Verdeans were part of the U.S. armed forces in World War I and World War II, he added.

The ambassador's visit was one of the several official calls Cmdr. Lindstedt made and received. "These people have awaited your arrival for a while," added Penner. "The embassy received a lot of calls to host the captain. I only wish your stay could be longer." Cape Verde became an independent nation on July 5, 1978. Until granted independence, the country was a Portuguese colony. The new government of Cape Verde is working hard to improve the standard of living for all Cape Verdeans. "They are developing, getting more trade and more currency," continued Penner. "Just last year several new hotels opened."

In addition to the ambassador's luncheon, the ship hosted Cmdr. Amancio Lopez, commander of the Cape Verde Navy, and his flag deputy Subtente Amante Da Rosa, and gave them a tour of the ship.

The commander remarked through an interpreter, "I have always wanted to see a U.S. warship and meet the crew. It was a pleasurable experience and one I will remember for a long time." Added Lopez, "This one ship is longer than our whole Navy."

WRITING STYLE, NEWS VERSUS OFFICIAL

Whether you compose hard-news releases or features, and whether you write for civilian or military sources, remember to write in "news" style, not the official Navy style appropriate for most of the items discussed elsewhere in this book. Here are a few pointers on news style.

Avoid Acronyms

Far too many military personnel send out press releases or articles full of unidentified acronyms. Proper news style is to use as few acronyms as possible, whether identified or unidentified. Even if a story is intended for a Navy or Marine Corps audience, spell out all acronyms the first time—but then use them as little as possible. To save space (and help the reader), look for a generic term instead: for example, "the wing" instead of COMASWWINGPAC or "the training group" in place of FASOTRAGRUPAC.

Remember, many readers may have just joined the service, and others are civilian employees or dependents with little service knowledge. Even among longtime service members, will a submariner understand aviation acronyms? Will an infantry Marine follow Pentagon budget terms?

Translate Jargon

Of course, avoid Navy and Marine Corps jargon and technical terms like "displacement." If you must use a specialized term, include a clear definition or explanation. Realize that even Navy ratings can be jargon, too. What would the public think a "boatswain's mate" or a "quartermaster" does? It's best to call that person BM2 Jones or QMC Patel and briefly explain the job the individual does. Better yet, quote QMC Patel explaining her own rating.

Use Media Abbreviations for Military Ranks and Other Titles

Another characteristic of news style is that it uses special abbreviations for military ranks—"Rear Adm." instead of "RADM," for example, and "Lt. Cmdr." instead of "LCDR." The list appears in *The Associated Press Stylebook*.

Capitalize Less

Newswriting capitalizes less frequently than other types of official documents. When writing for a newspaper or magazine, lowercase such occupational titles as "commanding officer" and capitalize the rank or job of "captain" only when immediately preceding a name—for example, Captain Evans. Unlike standard Navy usage, news style does not write the names of ships in all caps, underlines, or italics.

Quote Sparingly, Attribute Always

Besides adopting slightly different styles, news writers have to pay particular attention to the use of quotations. Quote people directly for unique, surprising, or striking statements or for important comments by important officials. At other times, however, it's best not to quote directly, such as when you're dealing with simple, factual material or when the exact wording of the quotation does nothing to enliven the story.

Proper attribution—stating the source of information—is vitally important for quotations and specific information obtained by sources, even if not quoted directly is attribution.

Of course, you need not cite a source for common knowledge, but for information that is subject to argument or involves policy, you must keep yourself out of trouble by citing specific sources. Use "he said" or the equivalent at least once per paragraph when citing such data. For variety, insert such attribution between sentences or phrases.

Incidentally, do not be reluctant to use that specific word "said" (as in "he said," "she said," or "they said"). You might occasionally substitute an equally neutral word like "remarked," "stated," "added," or "commented." But be careful with words such as "charged," "asserted," or "argued" that can imply a particular emotional element on the part of the person quoted that you do not intend. Similarly, if you report that a CO "claimed" or "maintained" something, the reader may infer you don't quite believe him or her. Be sensitive to the nuances of the words you use.

Direct quotations always need attribution, and any direct quotes should of course be absolutely accurate. Make sure you know exactly what was said, especially when quoting directly. You'll be surprised at how much trouble a slight misquoting can cause and how much it can upset the person misquoted.

Be aware of one slight exception to this rule of accuracy: If a speaker uses improper English (for instance, sentence fragments; bad grammar; lots of "ands," "ifs," or "buts"; or subject-verb agreement errors), common practice is to clean up those errors in the quotation for the speaker. Indeed, some people even let you make up comments for them. Just be sure to show the corrected quotation to the person before publication.

Below are some examples of various ways to present and attribute the same quotation. Besides noting the method of attribution, also pay careful attention to the punctuation and capitalization in the various ways of presenting the same quotation.

The Basic Information

> CAPT Charles B. Stevens made the following statement: "For 20 years we have provided the Navy with the best possible communications officers and radiomen."

With Direct Attribution in a Complete Sentence

> "For 20 years we have provided the Navy with the best possible communications officers and radiomen," the captain said.

or

> Capt. Charles B. Stevens, commanding officer of the communications school, said, "For 20 years we have provided the Navy with the best possible communications officers and radiomen."

or

> "For 20 years," the captain said, "we have provided the Navy with the best possible communications officers and radiomen."

With Direct Attribution, but Using Just Part of the Quote

> The captain said the school has "provided the Navy with the best possible communications officers and radiomen."

or

> The school has provided the Navy with "the best possible officers and radiomen," the captain said.

Using Indirect Quotation (minor rewriting into your own words with the same meaning)

> The communications school has aimed at providing top communication officers and radiomen for the Navy, the captain said.

or

> The captain said the school has always aimed at providing the Navy with the best communications officers and radiomen.

Further Style Guidance

For further guidance, see the most current edition of *The Associated Press Stylebook*, the primary guide that Navy and Marine Corps journalists and PAOs use for news style. That guide is the best quick reference on the style of newswriting available; it also has excellent advice on punctuation, word usage, capitalization in newswriting, and many other newswriting matters.

Additional, military-specific guidance for Navy and Marine Corps writers appears in the *U.S. Navy Style Guide*, the most recent edition provided in Appendix B of this book. Although this pamphlet provides much excellent guidance, it is not an official guide for general naval style as its title suggests. For instance, its overall guidance that "Navy editors and writers should follow the most recent edition of the *Associated Press Stylebook*" simply is not standard for internal Navy correspondence, directives, and other administrative documents. Many standard rules for those types of writing appear in the *Naval Correspondence Manual* (see chapter 2 of this book).

PUBLISHING THE STORY

Obviously, writing a story does no good if it doesn't get to the right outlets or doesn't get there in time. Your story will likely get only as far as a well-considered distribution list. Determine the public you are targeting and release information to the media that service that public.

Besides the local daily and weekly press (including base newspapers), don't overlook radio and TV outlets. Many press releases compete at the editorial offices of service-wide publications. When considering whether to send an article to such official magazines as *All Hands*, *Naval Services FamilyLine* (formerly *Navy Wifeline*), *Leatherneck*, or the commercial newspaper *Navy Times*, you need to ask yourself, "If I were in the Indian Ocean on deployment and saw this story appear in a publication, would it whet my interest?" If so, send it in. When submitting a press release or article, consider the audience and context for the publication. Taking special care to craft a release for a particular publication, group of publications, or geographic area can sometimes make the difference in actually placing your article.

INCLUDING AND CHOOSING PHOTOS

Send photographs with a story whenever possible. Photos (especially good quality pictures) will often convince an editor to publish a story.

Remember that in public-affairs photography, the individual is the important subject. Strive for an identity first, then the job, equipment, or background of the individual can usually appear in a photo without overshadowing the subject.

On the other hand, editors always look for pictures full of action and interest. Stress to your photographer that you want action shots. Instead of the ship's Sailor of the year receiving a certificate from the CO, have your photographer shoot the Sailor on the job.

Before you release a photo, ask yourself if anything in it detracts from the main subject. Does equipment or heavy shadow obstruct the faces? Could the photo be taken from a better angle? Is it in poor taste? Is it unflattering to the subject? If the answer to any of these questions is yes, throw it out and have it reshot. Not having a photo at all is better than using one of poor quality.

Be especially careful that the service members depicted present a positive image of the military, including its grooming and uniform standards. Any success you feel from having a story published will quickly disappear when the CO notices one of his Sailors has appeared before the world with an inch of hair over his collar.

For best effect (and likelihood of publication), avoid the following in your photos:

- flag poles, trees, and other objects appearing to stick out of a subject's head
- people's limbs cut off at unnatural spots
- idle hands—don't let them hang motionless at the subject's sides
- people staring at the camera
- dark backgrounds, especially when the subjects have dark hair
- still and formal poses, or "line-up" shots
- people just shaking hands; try to capture action and the unusual in your photos

Follow current public-affairs guidance on submitting digital photos. When possible, include a variety of views. Make sure to include a caption just under the photo (or taped to the back of a hardcopy submission). This should include the full names (and ranks/rates or titles, if appropriate) of depicted individuals and explain what is going on in the image. Give the photographer a photo credit, too. Follow this example:

> SN Mark W. Adams catches up on some damage-control maintenance while USS Bunker Hill (CG 52) is in port. A leading seaman in the ship's first division, Adams often performs duties required of a petty officer third or second class. (Photo by JO2 Mark Murphy)

FINAL POINTERS

Short items in magazines or newspapers often have surprising effect—a top-notch paragraph by itself may catch the eye, as can a striking photo with a deep caption (long cutline). Remember that the release itself should be easily readable—stay away from small or ornate typeface on your releases.

When you are deployed, base papers back home will be glad to get your copy. Friends and families will be delighted to see what their Sailors and Marines are doing. Indeed, for them, getting information published when you're deployed is probably more important than when you're at your home station.

Identify all the people in your stories (full rank or title, first name, middle initial, and last name), and be sure the names are accurate. Some commands insist that the CO's name be in the story you write, along with the full name of the command. On the other hand, remember that people, not ranks, make news. A lieutenant has no more news value than a seaman. Don't feel you have to mention the division officer and chief of every seaman you talk to. If you use quotes to enliven your story (not a bad idea), don't feel you must restrict your quotes to the CO or others in authority. The Sailor will often have something of even more interest to say.

Finally, make sure the appropriate officials have reviewed and approved your story before it goes anywhere. At a minimum, ask the PAO who should see your story; if there is no PAO, consult your XO for releasing guidance.

Resources for News and Social Media

For help in newswriting, become familiar with the following pertinent resources if you aren't already.

SECNAVINST 5720.44. The basic PAO regulation for both the Navy and Marine Corps.

The Associated Press Guide to News Writing, 4th ed., by Rene T. Cappon (Peterson's, 2019). Rather than a dictionary, this writing guidebook is especially good on newswriting.

The Associated Press Stylebook, 56th Edition: 2022–2024 (Basic Books, 2022). This book is the standard guide to news style for most American magazines and newspapers. It is also the universally accepted PAO style guide. Note, however, that U.S. Navy public-affairs officials have outlined some exceptions for Navy journalist-specific usage in the *U.S. Navy Style Guide*.

U.S. Marine Corps 2021 Social Media Handbook. A free PDF booklet available online, published by the U.S. Marine Corps Communications Strategy and Operations (COMMSTRAT) team.

Navy Social Media Handbook 2023. A free PDF booklet available online, published by the Navy Office of Information.

U.S. Navy Style Guide. This short pamphlet, included herein as appendix B, discussing a variety of matters of word usage (for example, when to use "aboard" versus "on board"), how to abbreviate ships' names, journalism abbreviations for Navy and Marine Corps ranks, and other information for writers. Published by the Office of the Chief of Navy Information, this guide pertains mainly to Navy social media and newswriting.

Writing for Social Media, by Carrie Marshall (BCS Learning, 2018).

In addition to the above sources, look for fleet instructions at your command. Almost every fleet or type commander issues local written guidance on newswriting (and has full-time PAOs who will be glad to give you advice and additional guidance).

11

The Professional Article

"*Professional Navy and Marine Corps articles allow the wider community to benefit from the unique expertise, perspective, and voice of the very Sailors and Marines who form the core of the sea services.*"

—RETIRED LIEUTENANT COMMANDER AND WRITING CONSULTANT

EXECUTIVE SUMMARY

Military professionals can offer their peers, seniors, and juniors fresh perspective on issues, leadership, technology, and other areas by publishing articles in professional journals. Such articles are less academic than school papers and more engaging than navy reports, but they might begin as other types of writing. A thoughtful planning process to determine the best focus and research what you want to say is essential to a strong article, as are careful structure, vivid examples, and a thorough revision process. Several professional journals related to the sea services provide a forum for Navy and Marine Corps writers to present their ideas to their colleagues.

WHY WRITE?

"*Stake out intellectual territory. Put forward a clear, forceful point of view. Leave no doubt of what you think.*"

—CAPT JOHN BYRON, NAVY,
"TEN COMMANDMENTS FOR *PROCEEDINGS* WRITERS"

So you're thinking of writing an article for a professional journal. Perhaps your spouse has remarked, as one commander did: "Why grouse so much to me? Why not write for *Proceedings* if you're so bothered by this?" Or maybe a shipmate has commented on one of your notions: "You know, that's an excellent idea, and I've never seen it spelled out anywhere. You ought to write it up." With such encouragement, should you take some extra hours to dig into this subject and write an article? The time and effort involved are a big sacrifice. You might wonder whether it will be worth it.

Of course, you'd like to see your name in print—always a motive for authors. Then there's the remote possibility that writing this article would do your career some good. Someone up the line might see it and think, "There='s someone who might make a good administrative assistant, speechwriter, or aide." Being recognized as a strong writer can open up many opportunities both in and out of the service. A pattern of authorship can have even larger consequences. Reportedly, Lieutenant Colonel Rommel "wrote his way to the command of a Panzer division and an eventual field marshal's baton with his Infantry Attacks" (Lt. Col. J. W Hammond Jr., "The Responsibility to Write," *Marine Corps Gazette*, January 1970, p. 28).

Yet some service members see little career incentive for publishing professionally. In fact, a number of officers commented about the dangers of publishing with warnings, with one warning: "Be careful when publishing. It's hard for officials to look at what you write with objectivity." Another stated, "You can say almost anything—but get it clear in your own mind the price you may have to pay." Comments like these suggest that you should understand the main reason for writing for publication. These officers didn't write articles primarily for personal prestige, career advancement, or monetary reward. Instead, they wrote for the good of the profession.

Why We Need Professional Journals

Most professions have journals that offer an exchange of knowledge, ideas, and opinions for those in that field. The military needs professional argument and open forums as much as, if not more than, other professions. Why not just live by official doctrine and official reports? By their very nature, official reports, reviewed and approved by the chain of command, tend to limit expression of views.

You might think that senior officers want to hear only supportive viewpoints. On the contrary, many are insulated from criticism by their positions and therefore truly need to hear some unofficial views on occasion. As Marine officer Gordon D. Batcheller pointed out in his essay "For the Sake of the Corps, Write!," published in the February 1984 issue of *Marine Corps Gazette*: "This [review process] has an inhibiting effect on all participants. . . . Critical candor becomes the first casualty." He went on to argue that because reports are usually staff efforts signed by committees, and because the signers become liable for the product, putting critical views forward is difficult: "Something you may feel with great conviction may not be shared by the officer who has to sign the document. You may be reluctant to ask him to buy into your misery; he may be unwilling to do so." As a result, the official information received at the top of the chain of command may be limited in value. As ADM Hyman Rickover once commented, "Always rely on the chain of command to transmit and implement your instructions; but if you rely on the chain of command for your information about what's going on, you're dead." One place to hear forthright personal opinions is in professional journals.

Not only commanders but all Navy and Marine Corps professionals need to have as informed a perspective as possible. True, we all read official publications that help us stay informed of late developments. But unofficial sources help greatly here, too. Most naval personnel at sea would know very little about larger developments in the services without such forums as U.S. Naval Institute *Proceedings*, *Marine Corps Gazette*, *Navy Times*, and *Naval War College Review.*

Informational articles in such forums keep readers abreast of new developments in hardware, strategy, tactics, logistics, and many personnel-related topics. Journals publish persuasive articles to plant ideas, open discussion, and help readers think of their jobs in new ways. Some pieces are "seminal" articles—they initiate whole programs and courses of action. Others do good work by "nudging" policy or thought. As Frank Uhlig, longtime publisher of the *Naval War College Review*, pointed out, "Even if you don't convince your readers immediately, you get readers to begin thinking about what you're thinking about."

So do your service a favor and put your good ideas or experience in print. How to proceed? The following pages offer our guidance gleaned from conversations with many military authors and editors and from published articles on this subject.

Like most writing, publishing an article involves a process with distinct steps. These do not have to follow a set order, but the writing process will go more smoothly if you don't try to tackle all of them at once. Here is a suggested process for getting your idea from your head to the pages of a professional journal or magazine.

FIGURE 11.1. *Opportunities for New Authors to Write*

Should you hesitate to write because of lack of seniority? Absolutely not. Most military journals—and particularly *Proceedings* and the *Gazette*—actively encourage junior authors and go out of their way to help those with promising ideas to improve their articles. They open their pages to enlisted service members, civil servants, and young officers. They even hold contests to stimulate new authors to try their hands at writing essays.

If anything, because of editors' encouragement of junior authors, someone who is not senior will have a marked advantage in getting an article published, especially the very first one. After that, you're an old hand, and will have to sweat and struggle like the rest of us.

Planning and Drafting

DECIDE ON YOUR FOCUS

The basic question is one of focus. First, decide whether you know something that your reader doesn't and that is significant or interesting enough to attract attention. Writing well depends above all on having something to say.

One good piece of advice to authors is to concentrate on a specific area—limit your subject. Don't try to refight the Gulf War in two thousand words or attempt to explain the Navy and Marine Corps budget process even in five thousand. Not only are such subjects impossible to cover in such a short space, but narrowing the subject to something you really are an expert in will help you to speak authoritatively. Stay in your area of expertise and write on a subject with which you're familiar.

That doesn't mean you have to have twenty years of experience before you write. As a senior military-journal editor pointed out, midshipmen sometimes have held their own with officers ten years their senior by writing about what it's like to be led by good or bad leaders or by revealing their thoughts upon first stepping into leadership roles. The midshipmen could speak with credibility and sharpness when they stayed close to their own certain knowledge. Of course, as you gain experience, you can range much further afield. But new authors should start with what is close to home.

CONSIDER WHAT AND WHERE TO PUBLISH

"From my own limited perspective, Proceedings *has been the only vehicle consistently available to articulate the truths about the limitations in our medical readiness posture. In fact, there is no other avenue for collegial discourse among professionals about our observations and suggestions short of surreptitious calls to 'Hot Lines,' abbreviated letters to the editor in* Navy Times, *or going through the stifling point paper process via chains of command."*

—CAPT ARTHUR M. SMITH, MC, UNITED STATES NAVAL RESERVE,
LETTER TO THE EDITOR, *PROCEEDINGS* (JANUARY 1996)

Having selected a subject, consider the kind of forum for which you're trying to write. Is what you have to say suited to a long informational article, or will it be more of an opinion piece? Which journals publish the type of piece you want to write? Locate at least three or four possibilities and read some sample articles to get a sense for their style and tone. Articles should be available online, or the journals may even send you older copies. You can also usually find submission guidelines online or inside the cover of any issue of the journal.

Analyze several back issues, taking a careful look at what each periodical publishes, especially the kind of articles it has put out recently. What are their style, length, and contents? For whom do they seem to be written? Not only will such analysis help you decide whether your article is appropriate for that journal, but it may also give you several ideas on good methods of support, ways to begin and end, the best use of anecdotes and quotations, and so on. Remember that a journal normally has several different sections. *Proceedings*, for instance, has Professional Notes and Nobody Asked Me, But, as well as feature articles. The *Marine Corps Gazette* is similar: each issue has several sections, and besides full-length articles of about 2,000–5,000 words, it publishes shorter ones (under Ideas and Issues) of 750–1,500 words. Some of the latter involve opinion or argument, while others are professional notes (for example, Strategy and Tactics or Weapons and Equipment) or short historical vignettes. Typically, the editorial-review process differs for different sections of a journal; getting a piece into the shorter sections is sometimes a bit easier.

Letters to the editor, by the way, can add much to the professional perspective. The *Marine Corps Gazette* submission guidance (easily available online) encourages letters that "correct factual mistakes, reinforce ideas, outline opposing points of view," and identify additional needed considerations. The guidance concludes with, "the best letters are sharply focused on one or two specific points."

The end of this chapter lists a few publications to which you might consider submitting your writing.

RESEARCH YOUR TOPIC

Of course, you must have excellent support for whatever kind of article you write. Your reader rightly expects you to have rich and detailed knowledge or experience assets that make you worth listening to.

Regardless of how much you think you already know about your subject, become familiar with its wider context by reading published sources, including past articles related to it. Besides giving you more information and assuring you of the quality of your ideas, these sources will help you orient your article in the current professional discussion. You don't want to submit a piece that says the same thing as an article a year or two ago. And if your article presents a view that differs greatly from an earlier work on the subject, you will gain credibility if you can reference it and address the points of disagreement.

You may find that you need to do additional research, or you may conclude that you already have sufficient arguments and data at hand. Whether you need to do further research is a judgment call on your part. The test of whether you have enough to present your topic fully is often the first draft.

CONVERTING A PAPER INTO AN ARTICLE

Many professional military articles originate in a classroom, but what makes a good student paper is not what makes a good article. The paper typically has a specific instructor as the intended audience (an expert who is already interested in the topic), and part of its purpose is to demonstrate the student's research. Thus, the paper overflows with buzzwords and stumbles with weighty footnotes. Many journal readers will have never so much as thought about the subject before, so the author must spread a much wider net by cutting back mere documentation and enlivening crucial support.

Typically, those who have written a good paper at Newport, Quantico, Monterey, a civilian school, or some other academic institution must describe their subjects much more simply (and visually) than they did in the original essay. They must also change the terms they use; the terminology and acronyms that work for experts are often gibberish to others.

In fact, so specialized has military terminology become that when writing for a broad military audience, an author does best to assume that readers aren't in the service at all. As Frank Uhlig of the *Naval War College Review* once put it: "Assume that the people reading your article are intelligent and interested laymen. A destroyer officer and a fighter pilot both belong to the same navy, but neither is likely to know very much about the other's business. The same is true if you are aiming at people in other services, or civilians, only more so."

Adding intriguing introductions, simpler and fuller explanations, descriptive language, simplified terms, and striking conclusions can transform a classroom paper into a journal article. Readers can always put the magazine down when what you have written does not engage them, so you must work for relevance, interest, and even charm. The later section "Revise for Structure" provides examples of openings (or "leads"), illustrations, and conclusions.

DRAFT THE ARTICLE—GET THE CONTENT DOWN

Content is, of course, the key to any article. Journal editors have great patience with authors who have dynamic ideas, even if those writers need help polishing their style. On the other hand, if your content is weak, whatever catchy introductions and other stylistic dressing-up you add won't help. One prominent military-journal editor complained about bad content even in articles submitted from war colleges. Though an article by definition is conceptual, it must at least nod toward real-world issues like cost, manpower, priorities, and others.

Therefore, when writing a professional article, concentrate on the basic idea and its proof (if yours is an argumentative essay) or on a sharp presentation of content and data (if it is an informative piece). Here are some of the questions that *Proceedings* editors have often asked of an article that crosses their desks:

- Is the subject relevant?
- Is its focus sharp and the topic limited?
- Is the topic timely?
- Does the author know what he or she is talking about?
- Is the logic sound?
- Has the research been thorough?
- Is the thought sequence easy to follow?
- Do the main points emerge clearly from the details?
- Are the explanations clear and convincing?
- Are the details interesting?
- Has the writer had experts criticize this paper?

Editors can fix the grammar, and they can help an author with an introduction, a closing, and the whole style of a piece. But a miner mines raw material only to recover some precious metal within. Make sure your essay has good substance to begin with.

CONSIDER VISUALS

Visual elements such as photographs, charts, tables, maps, and diagrams can make your argument much more readable and interesting. They can also incline an editorial board in your favor, assuming the visuals are clear. Sometimes a staff artist can enhance a rough sketch. Of course, think in terms of lively illustrations meant for magazines, not dull charts from a training manual. Although not all articles include illustrative material, it may help sell your piece. If you have an idea for artwork but can't get it yourself, let the editors know. A magazine (particularly *Proceedings* or *Marine Corps Gazette*) can often illustrate your article with photographs from its own files.

REVISING

"Edit ruthlessly."

—CAPT PAUL RYAN, NAVY, "HOW TO WRITE FOR *PROCEEDINGS*"

REVISE FOR CONTENT

After getting a good initial draft down, work it over again. (Wait a couple of days first, and you'll be surprised how different it all looks.) Flesh out your ideas and generally rework the piece. You may find you need to modify your thesis, your central assertion, or your idea. Much of your effort will be adding details, examples, and illustrations while also looking at your logic. All of this work can be tedious, but if your first draft was successful, you can often revise a section at a time until you've substantially revised the whole essay.

When deciding how much to say, follow this general rule: Don't say everything you know about the subject, but do say everything the reader needs to know in order to understand and to buy your main point.

Here are some other tests for the content of your article, once you're in the revision stage:

- Is your idea sound? Is it so clearly presented that it seems self-evident to the reader?

- In an argumentative article, have you taken a strong stand, adopting a definite point of view?

- Do you have sufficient support? Do you justify your assertions and give evidence for any generalizations? Evidence can be facts, descriptions, graphs, tables, anecdotes, and many other things. Of course, double-check all your facts. Remember, yours is the final responsibility for what you say. Unlike scholarly journals, magazines will usually rely on your subject-matter expertise; they do not necessarily check the basic accuracy of all that they publish.

- Have you included quotations (if available and relevant)? The words of an authority on your topic can add both interest and credibility to your article. But don't clutter your manuscript with long quotes that only restate what you are trying to prove. Look for the telling comment, the penetrating phrase, or the expression that precisely captures the issue at hand.

- Do ideas carry weight because of the authority of the person who uttered them? Remember that even if you do not quote someone directly, you must still cite them as a source for the idea or information you use.

- Have you documented all special sources and data? Editors are wary of any unsupported assertions.

- Are your clear? Have you considered the specific audience you are writing for and its level of understanding? Make sure you explain difficult concepts and technical terms.

- Have you clarified your argument's relevance? That is, have you made clear why your argument matters and what its consequences are in terms with which your readers can identify?

- Have you anticipated and refuted any opponents' likely arguments?
- Finally, if you have criticized a policy or program, have you also offered constructive recommendations? Avoid criticizing a policy or procedure without suggesting an alternative.

REVISE FOR STRUCTURE

As discussed with so many other types of writing in this book, the organization of ideas has a tremendous influence on how clearly your overall message comes across. Some written (or spoken) documents can work well with a variety of structures; others only work in a certain order. Structure—used synonymously with "organization" here—is the order you arrange the parts of your article. Should you open with an interesting story or save it for the middle of your article? What order should you arrange your main ideas? There is rarely a right answer to these kinds of structural questions, but here are some suggestions that should help.

- Place your thesis, your central idea, in sentence form, near the beginning, ideally somewhere in the first or second paragraph.
- Put your most important point first. Generally, the most effective arrangement of conceptually parallel ideas, such as three or four main points, is to provide the most valuable or most interesting point first and continue in order of decreasing importance. Some people find this idea counterintuitive because they want to tell more of a story that gradually builds to a climax, like the plot of a good movie or novel. The problem with this approach is that your readers may not stay with you all the way to the end. Instead, give away the main idea in your first paragraph (the BLUF approach).
- Use headings and even subheads every few paragraphs, depending on the content. Subdivision will help keep the reader oriented and can make it easier for those who are skimming to become interested in reading the whole piece.
- Note that the above recommendations are similar to those in the discussion of organization for any document as provided in chapter 1.

Openings, or "Leads"

A good beginning not only attracts the reader's attention but also draws the person immediately into important material. A "lead" (similar to a lead—or "lede"—in a news article, as discussed in chapter 10) can be a sentence, a phrase, or a paragraph. When describing new programs or recent developments, you can sometimes count on ready reader interest and therefore keep the lead fairly brief. On the other hand, the lead for a feature article will occasionally extend over several paragraphs—more like a full introductory story. Be careful not to lose momentum and thus lose your readers' interest. Craft the lead to draw your audience quickly into the meat of the article. The following examples illustrate a few ways to do so effectively.

This opening captures the attention with current details readers can relate to:

COVID-19 brought with it the need to adapt to new workplace environments, primarily online. While the Navy has made great strides in its capability to telework since the start of the pandemic, it initially was woefully unprepared. The service quickly realized the 240,000 simultaneous connections for Outlook Web Access and 40,000 for a virtual private network (VPN) could not accommodate the demands of its approximately 619,000 personnel.

— LCDR Eric Zilberman, USN, and Cynthia A. Kidd, "Abandon NMCI for Commercial Solutions," U.S. Naval Institute *Proceedings* (December 2020): 14.

You can also begin with a striking statement that leads to the central issue:

Are Army infantry soldiers more important than Marines as human beings or as components of U.S. defense strategy? If this strikes you as a preposterous and insulting notion in 2006, more than six decades after Iwo Jima, consider the fact that Marines still do not have a dedicated medevac helicopter—a nicety of war that is all about saving limbs and lives and that Soldiers have had for many years.

— CAPT Michael Vengrow, MC, USN, "Saving Limbs and Lives," U.S. Naval Institute *Proceedings* (February 2007): 20.

Here, the first sentence quickly focuses the article:

The Marine Corps loses the talent of innovative and capable Marines because of the poor leadership they experience during the first four years. After their first contract, they're gone and are never coming back.

— CPL Michael K. Adams, USMC, "Leadership and the Private's Agenda," U.S. Naval Institute *Proceedings* (July 2006): 86.

This timeless, attention-drawing anecdote draws readers in with narrative to introduce the topic:

A scientist and an engineer were put in a room across from a bag of gold. They were told they could have the bag when they moved across the room.

They could go as fast as they wanted provided they went no more than half the remaining distance with each move. The scientist, recognizing the impossibility of the situation, left the room. The engineer, on the other hand, moved across the room until he was within arm's reach of the bag. At this point, he declared that he had gone far enough for an engineering approximation, grabbed the bag, and left.

In developing its weapons, the Navy's goal frequently is to build a state-of-the-art system on the cutting edge of technology, rather than a simpler system that is good enough to get the job done. The result is that developmental work is always halfway finished, and its goals are never reached.

— LCDR Eric Johns, USN, "Perfect Is the Enemy of Good Enough," U.S. Naval Institute *Proceedings* (October 1988): 37.

Illustrations and Examples

Making bare facts or general statements meaningful can require special methods such as visualization, comparison, exemplification, and others. The possibilities are countless, but the techniques below are common.

Support assertions with examples. Details that support your general claims will both illustrate what you mean and tend to convince—by accumulation if you have several.

> The LCS [littoral combat ship] demands sailors who are technically proficient not only in their trades, but also in a wide range of collateral duties. For example, a gifted information systems technician who leads a communications shack also is expected to serve as the command career counselor; facilitate morale, welfare, and recreation events; and manage the command resiliency team.
>
> — LT John Albani, USN, "LCS Sailors Will Lead the Fleet Forward," U.S. Naval Institute *Proceedings* (January 2021): 29.

Use statistics to make your point clear and persuasive. Careful use of these and other data can make a concept much more clear. Make your figures comprehensible to your reader with comparisons, multiplications, or other such manipulation.

> SpaceX's ability to launch payloads for roughly $2,700 per kilogram for low-Earth orbit and $7,500 a kilogram for geostationary transfer has revolutionized satellite possibilities. Competitive pricing has resulted in a proliferation of systems—more than 300 satellite launches in 2019, compared with roughly 65 a decade earlier. In fact, the next ten years will see SpaceX,

OneWeb, Amazon, and others fielding megaconstellations comprising hundreds and thousands of satellites.

— LCDR Mark Wess, USN, "ASAT Goes Cyber," U.S. Naval Institute *Proceedings* (February 2021): 46.

Note how the superintendent of the Naval Academy used two sets of statistics in this article from the 1990s to refute a widespread misconception:

Universal "Truth" Number Two:

Service academy graduates do not stay longer.

This perception does not come from any data officials at the service academies are familiar with. Retention rates for Naval Academy graduates exceed all other officer accession sources at every major career decision point. For example, to produce 40 career-designated officers—those with at least ten years of service and selected for lieutenant commander—requires an initial accession of 100 Naval Academy graduates, 140 from NROTC, and 153 from Officer Candidate School (OCS). Furthermore, the last class that was tracked to reach the 20-year point had a retention rate of 41% for the Naval Academy, 24% for NROTC, and 21% for OCS. So that universal truth has absolutely no factual basis.

— ADM Charles R. Larson, USN, "Service Academies—Critical to Our Future," U.S. Naval Institute *Proceedings* (October 1995): 34.

You can also use bullet format in articles just as you can in other documents, as in the following argument against the multitude of rules governing captain's mast in the Coast Guard. Notice how the author supports his thesis with multiple details.

For example, if a Coast Guard, non-rated enlisted member is awarded non-judicial punishment, the following administrative measures are required:

- Member is no longer eligible for transfer without CG Personnel Command approval.
- Supervisor must prepare a special, adverse evaluation.
- Supervisor must assign an "unsatisfactory" mark in conduct on the member's evaluation, with automatic loss of Good Conduct eligibility.
- Member cannot be advanced during the remainder of the marking period.
- Member is ineligible to apply for officer commissioning programs for 36 months.

· Member must be removed from any "A" school waiting list and is ineligible to reapply for 6 months.

· Member is ineligible for special duty assignments for 4 years . . . (and the list goes on).

— CDR Kevin E. Lunday, USCG, "Repeal the 16-Pound Sledgehammer," U.S. Naval Institute *Proceedings* (February 2007): 41.

Closings, or "Sign-Offs"

A good closing, or "sign-off," to an article will usually lead back in some way either to the heart of the article or to some special high point in it. Like a strong conclusion to an academic paper, the closing of an article can point readers to implications, applications, or even recommendations arising from the main discussion. A closing can briefly summarize the central idea so that it can more clearly connect that idea to whatever the author wants the readers to do with it or think about it. Whatever the approach, the best closing provides closure, releasing the reader's attention gracefully but with a final flick of the wrist. At the very end (the last sentence or words), you might hark back to the introduction, cite a historical quotation, or make some kind of appeal.

This closing reminds readers of the central thesis, elaborates with a quotation by an authority (a former CNO), and then challenges readers:

> When Navy and Department of Defense leaders speak of the submarine force's competitive advantage, they also warn that it cannot be taken for granted. As former CNO Admiral Jonathan Greenert wrote, the undersea is "the one domain in which the United States has clear maritime superiority— but this superiority will not go unchallenged." Each day, we must heed those warnings and work to protect, preserve, and increase our edge to ensure the nation can rely on the Navy's advantage in the undersea.
>
> — VADM Daryl Caudle, USN, "Sustaining the Submarine Force's Competitive Edge," U.S. Naval Institute *Proceedings* (October 2020): 25.

This closing to an article from the early 1980s recaps briefly and then invites readers to apply the point of the article to themselves through a quotation:

> The story of our POW experience in Vietnam, accepting a modicum of failure, is one of undaunted, unremitting courage. Even a fleeting profile of POW opinion demonstrates that the Code of Conduct proved to be sound doctrine. Capt. Jim Mulligan put the Code in final perspective:

It can't answer everything, but it sets the rules. If you don't have any moral guts or personal integrity, the Code is not going to give them to you. But most military people have them someplace inside, and the Code of Conduct brings them out.

— MAJ Terrence P. Murray, USMC, "Code of Conduct—A Sound Doctrine," *Marine Corps Gazette*, December 1983, 62.

Here is another example of closing with a call to action:

A system that requires individual Sailors to cover governmental obligations until it is convenient for the government to pay is fundamentally flawed and unreasonable. We owe it to our Sailors to investigate other options and make the travel process more reasonable and less onerous to those least able to bear the burden. For our Sailors' sake, the DoD travel card must go.

— LCDR Chris Davis, USN, "The DoD Travel Card Must Go," U.S. Naval Institute *Proceedings* (February 2007): 67.

REVISE YOUR ARTICLE FOR STYLE

There are many ways to say the same thing. Once satisfied with the content and structure, work on your style—word choice, tone, and level of technical vocabulary. This step is different from proofreading for errors. Review the entire article for

---

FIGURE 11.2. *Tips on Writing Styles for Professional Articles*

- Is the writing vivid, or does it sound like a training manual or an official instruction? If the latter, try rewriting sentences to sound the way you speak (minus slang and curse words, of course). That approach will freshen up your language. Virtually no one talks like a Navy regulation or technical manual—so if you talk out your article, the writing will usually be clearer, simpler, and more natural.

- Do you use regular action verbs, or have you replaced them with cumbersome noun forms? For example, instead of "operationalization," say, "the process of putting a process into operation." Instead of "perform test and evaluation procedures," just say "test and evaluate."

- Have you used active voice for most sentences? Virtually all military journals mention this point. None of them wants all that impersonal and longwinded discourse so common to navalese. Especially avoid such phrasing as "it is recommended that," "the suggestion is made to," "the argument is offered that," and so on.

- Is your tone respectful of the Navy, of those you disagree with, and of those senior to you? You will actually be more persuasive if you write in a style that conveys respect for those whose position or views you critique.

- Do your sentences tend to use 12 words when 10 would convey the same idea?

FIGURE 11.2. *(continued)*

- Have you repeated yourself unnecessarily? Military authors tend to repeat themselves in their writing, and they often provide too much background and context. As a result, military journals often cut two or three pages from both ends of an article. Watch for long sentences and cut them back (an average of about 17 words per sentence is ideal, according to some experts). Keep the paragraphs short, too. Cut out all words, sentences, or paragraphs that are dull, repetitive, or unnecessary.

- Work on rhythm and phrasing, reading aloud to listen to your prose. Strive for exactly the right word.

- Finally, have you avoided jargon? Define all important terms. Avoid unnecessary abbreviations and acronyms. Define what acronyms you absolutely must use the first time you use them, as in tactical action officer (TAO), remotely piloted vehicle (RPV), or airborne early warning (AEW). Then, you can use these acronyms in succeeding sentences as long as they aren't several paragraphs apart. Vary subsequent references by interspersing other, simpler nouns ("the officer," "the vehicle," "the concept," and the like) to keep from overusing the acronym.

---

careful word choice, clear sentences, and other sentence-level selections. Although journals edit articles before publishing them, an editor is more likely to publish a polished article than a dull, awkward, and unclear text. Consider these aspects of your style as you review.

GET GOOD CRITIQUES, AND THEN REVISE AGAIN AND PROOFREAD

Once you have a good draft, get some good criticism. One frequent contributor to *Proceedings* has argued that you should find ten people who know a lot about your subject and have them tear your article apart. He always did. As a result, he found that his material sailed through editorial staffs—that made him look a lot smarter, too.

Maybe you can't get ten, but get at least a couple of people knowledgeable about your topic to review your work. And to make sure you are not speaking over your readers' heads, have some nonexperts read your article as well.

After you get their critiques, revise your essay once more. At this stage you should also proofread for grammar and punctuation. As with any other documents discussed in this guide, use spell-check and grammar check but make sure any changes you accept make sense.

PREPARING FOR SUBMISSION

GET THE REQUIRED SECURITY AND POLICY REVIEW

Normally, service members and DoD civilians must clear articles for security through their local public-affairs offices prior to submission and must include a

> *"The trick is to make your essay jump out of the stack of 90 or so that the*
> *staff readers (including me) must screen—with a grabby title, snappy motto,*
> *riveting lead paragraph, and challenging arguments. You want to grab the reader*
> *by the stacking swivel, move him deftly from Point A to Point B, and convince him*
> *that the trip was worth his time and effort."*
> —COL JOHN MILLER, UNITED STATES MARINE CORPS,
> FORMER MANAGING EDITOR, *PROCEEDINGS*

signed statement of clearance with the article. Although some journals will assume responsibility for a security review, usually the author must get the article cleared. Don't regard a security review as antagonistic; do your homework ahead of time.

Not only security but policy may be an issue. For one thing, be careful about using the privileged information your job gives you access to, even if it is unclassified. Such information may not be intended for public distribution. Wise authors make a practice of getting the material they use from some standard naval source and then citing that source in their articles.

Also, recognize that readers may have difficulty separating your own view from that of the office that you occupy. Even with a useful disclaimer on your part—"Opinions, conclusions, and recommendations expressed or implied within are solely those of the author and do not necessarily represent the views of [your office/department]"—people will often assume that the stance you take is official because of your official position. Talk to experienced authors around your duty station or community if you envision any problems in this regard.

Another difficult subject is how to handle legitimate dissent within such a hierarchical and honor-bound institution as the naval services. On this subject a senior officer once gave this evenhanded advice:

> Your oath of office doesn't cease when sitting at the typewriter. You have a professional obligation (1) not to reveal any classified information; (2) to be responsible, objective, and not harmful to the service; (3) to learn the current rules for getting articles cleared for policy as well as security; and (4) never to sandbag (i.e., surprise) your boss; send a courtesy copy to her or him when your article has been prepared for publication.

Regardless, remember that there will always be someone who doesn't like what you have to say; you cannot please everyone. A magazine editor once suggested: "Try to stay within the regs, but don't be slavish to them. If you run into review problems, contact the editor. Maybe he can help."

FINAL REVIEW

As a final review, check all figures, dates, names, titles, footnotes, quotations (especially quotations—and against the original source), and other material to make sure you have not made inadvertent errors. Also check grammar, mechanics, spelling, and punctuation one more time (a crucial step for anything you write).

FORMAT THE ARTICLE AND SUBMIT IT

Double-check the exact format required by the journal to which you're sending the article. As noted above, most journals describe on their websites what they look for, including format and number of words for different kinds of articles. Also, a journal will usually list its specific format requirements in each issue somewhere near the title page. Some journals require footnotes; others ask you to work documentation into your text (as done in this chapter). Technical journals will often ask you to submit an abstract—that is, a summary. Some journals will accept articles as attachments to emails.

In your submission, unless directed otherwise, include your name, rank, office, and phone at the beginning or end of your article—and number your pages.

For images, check with each magazine for whether they prefer black-and-white or color photographs and what type of file or format they want. If you are submitting print photos, identify pictured individuals on a separate sheet (writing on the back of a photo can damage it). For all illustrations, include a suggested caption, which should explain the illustration and relate it to the text as appropriate, but it shouldn't repeat the exact words of the text.

Remember that most publications ask authors not to submit material to more than one journal at a time; they regard multiple submissions as unprofessional. But check each journal's policy on this.

PUBLICATIONS

Consider these naval and other military journals when deciding where to submit your article. These are four major professional magazines or journals that provide forums for expression of opinion on the naval profession:

- U.S. Naval Institute *Proceedings*
- *Marine Corps Gazette*
- *Naval War College Review*
- *Naval History*

In addition, some official Navy or Marine Corps magazines to consider when submitting material include the following:

- *Leatherneck*
- *Naval Aviation News Submarine Review*
- *Seapower*
- *Surface Warfare*
- *Navy Civil Engineer*

Wherever you submit your article, be prepared to make further revisions as requested (or required) by the journal prior to publication. If one rejects your submission, try to get some feedback about why and submit it somewhere else. Many successful writers have persevered through many rejections on the way to getting their work published. And if you have something you are convinced others need to read, you do those readers-to-be a service by getting your ideas in circulated print.

Resources for Naval Writing

Some chapters in this book mention other reference works, Navy and Marine Corps instructions, and guides you can consult on specific areas of writing. The list below consolidates those lists and includes some additional resources you may find helpful. Note that publication and edition information are current as of 2023. If a new edition of any of the following is available, we recommend you use it instead.

Navy, Marine Corps, or other Government Publications and Instructions

Department of the Navy Correspondence Manual. SECNAV M-5216.5; also used by the Marine Corps.

Navy Office of Information. *Navy Social Media Handbook 2019.* https://www.csp.navy.mil/Portals/2/documents/downloads/navy-social-media-handbook-2019.pdf.

Navy Office of Information. *Navy Social Media Playbook for Leaders 2022.* https://media.defense.gov/2022/Jun/28/2003026453/-1/-1/0/SMP_LEADERS_FINAL.PDF.

U.S. Government Publishing Office (GPO) Style Manual. Washington, DC: GPO, 2016.

U.S. Marine Corps Communications Strategy and Operations (COMMSTRAT). *U.S. Marine Corps 2021 Social Media Handbook.* https://www.marines.mil/Portals/1/Docs/2021USMCSocialMediaHanbook.pdf.

Books on Writing and Style

The Associated Press Stylebook: 56th Edition, 2022–2024. New York: Basic Books, 2022; also available online as *AP Stylebook Online* (subscription required).

Brusaw, Charles T., Gerald J. Alred, and Walter E. Oliu. *Handbook of Technical Writing.* 12th ed. New York: St. Martin's, 2020.

Casagrande, June. *The Best Punctuation Book, Period: A Comprehensive Guide for Every Writer, Editor, Student, and Businessperson.* New York: Random House, 2014.

Eisenberg, Anne. *Guide to Technical Editing: Discussion, Dictionary, and Exercises.* New York: Oxford University Press, 1992.

Garner, Bryan A. *Garner's Modern English Usage.* 5th ed. New York: Oxford University Press, 2022.

Lannon, John M. and Laura J. Gurak, *Technical Communication.* 15th ed. New York: Pearson, 2019.

Markel, Mike. *Technical Communication.* 13th ed. New York: St. Martin's, 2020.

The Microsoft Manual of Style for Technical Publications. 4th ed. Redmond, WA: Microsoft, 2012.

Strunk, William, and E. B. White. *The Elements of Style.* 4th ed. New York: Pearson, 1999.

Online Resources

Air University Library Index of Military Periodicals (AULIMP). https://fairchild-mil.libguides.com/AULIMP.

Grammar Girl: Quick and Dirty Tips. https://www.quickanddirtytips.com/grammar-girl/.

Navy Writer. navywriter.com.

Plainlanguage.gov.

Purdue Online Writing Lab (OWL). College of Liberal Arts, Purdue University. https://owl.purdue.edu/owl/.

Resources for News and Social Media

Cappon, Rene T. *Associated Press Guide to News Writing.* 4th ed. Peterson's, 2019.

Marshall, Carrie. *Writing for Social Media.* BCS Learning, 2018.

SECNAVINST 5720.44.

In addition to the above sources, look for fleet instructions at your command. Almost every fleet or type commander issues local written guidance on news writing (and has full-time PAOs who will be glad to give you advice and additional guidance).

U.S. Navy Style Guide[1]

Navy editors and writers should follow the most recent edition of the *Associated Press Stylebook* except as noted in this *U.S. Navy Style Guide*, published by the U.S. Navy Office of Information (CHINFO).

"A" school—Use double quotes throughout a story. If included in a quote, use single quotes: *'A' school.*

abbreviations, **acronyms**—Upper case initialisms and acronyms. Spell out on first reference, with the abbreviation in parenthesis. Some acronyms, such as "NATO," can be used on first reference. Check the *AP Stylebook*.

- The individual augmentees (IAs) met May 5. All Sailors reporting to IA duty are invited to attend.

Other examples:

- BUMED—Bureau of Medicine and Surgery
- CIWS—close-in weapons system
- CNO—Chief of Naval Operations
- OCONUS—Outside Continental United States
- OPTEMPO—Operations Tempo, or Tempo of Operations
- RHIB—rigid-hull inflatable boat
- SECNAV—Secretary of the Navy

aboard/on board—Use "aboard" when referencing events taking place on a ship or aircraft. Use "on board" when discussing shore-based events.

- The crew is *aboard* the ship.
- The memorial ceremony was held *on board* Naval Station Norfolk.

---

1. Adapted from *U.S. Navy Style Guide*, version 16-2, Dec. 30, 2016. https://www.navyfitrep.com/wp-content/uploads/2017/02/US-Navy-Style-Guide.pdf. Accessed November 2023.

- Also, a Sailor is stationed "on," "at," "is serving with," or "is assigned to" a ship. A Sailor does not serve "in" a ship.
- A ship is "based at" or "homeported at" a specific place. A plane is "stationed at" or is "aboard" a ship; is "deployed with" or is "operating from" a ship. Squadrons are "stationed at" air stations. Air wings are "deployed with" ships.

active duty, **active-duty**—As a noun, two words: Navy personnel serve on *active duty*. As an adjective, hyphenate: All *active-duty* personnel must participate.

aircraft—Acceptable characterization of naval aviation platforms. Do not refer to military aircraft as "airplanes" or "planes."

aircraft designations—Always used as a letter(s) followed by a hyphen and number: SH-60B Sea Hawk or F/A-18E/F Super Hornet.

aircraft squadrons—Spell out full name of squadron on first reference. On second reference, use abbreviation and hyphenate.

- Strike Fighter Squadron (VFA) 97 deployed aboard USS *Carl Vinson* (CVN 70). During their deployment, VFA-97 maintained a perfect safety record.

aircrew, **aircrew member**—Per *Webster's*, one or two words.

air wing—Use as two words.

all hands, **all-hands**—Two words as noun: He called *all hands* to the meeting. Hyphenate as adjective/compound modifier: They attended the *all-hands* call.

Anchors Aweigh—not "Anchors Away"

anti-aircraft, **anti-submarine**—Hyphenate.

Arabian Gulf—Use instead of "Persian Gulf" per Commander, Naval Forces Central Command, U.S. 5th Fleet.

armed forces—Capitalize only as a proper name (*Armed Forces* Day), not as a noun (the *armed forces*) or adjective (an *armed-forces* member). It is lowercased unless part of a title or when preceded by "U.S.," as in U.S. *Armed Forces*.

attribution—Identify the source of reported information; especially objective and opinioned-based statements. Include context in which comment was made if it is not apparent.

- Use "said" with quotes. Do not use "says."
- See **quotation marks**.

battalion—Use numerals in unit names, spell out on first reference and abbreviate and hyphenate on second reference:

- Naval Mobile Construction Battalion (NMCB) 4; NMCB-4 (not NMCB FOUR)

battle group—Do not use *battle group*. Rather, use "carrier strike group" or "expeditionary strike group."

boat—Use to describe a submarine. Do not use to describe a ship.

boot camp—Use as two words.

burial at sea—Do not hyphenate.

call signs—Do not refer to individuals by *call signs*. Use full name and rank.

carrier strike group—Capitalize when used with the name of a ship. Precede name of strike group with "the."

- The *Harry S. Truman Carrier Strike Group* arrived in the U.S. 5th Fleet area of operations.

chaplain—*Chaplains* are identified as "CMDR John W. Smith, a Navy *chaplain*," in the first reference and as "*chaplain*" or by last name thereafter.

chief (select)—Use the service member's current rank: "Hospital Corpsman 1st Class Franklin Pierce will be promoted to chief petty officer next month." Do not use "select."

Chief of Naval Operations—Lowercase when referenced after an individual's name or when used alone.

chief petty officer—Applies to Navy or Coast Guard personnel in pay grade E-7. Lowercase when referenced after an individual's name or when used alone.

Chiefs Mess—Do not include an apostrophe.

cities/datelines—For cities that stand alone, use the list of datelines found in the *AP Stylebook*. Because of their strong Navy ties and frequent reference in stories, Norfolk, San Diego, and Pearl Harbor can stand alone, without states.

civilian titles—Use full name and title or job description on first reference. Capitalize the title or job description when it precedes an individual's name and do not use a comma to separate it from the name. Lowercase titles when they follow the name or when not accompanied by one.

- Assistant Secretary of the Navy (Manpower and Reserve Affairs) Franklin R. Parker holds an all-hands call.
- Davey Jones, Naval Station Norfolk historian, logs in artifacts.

Use last names only on second and all following references. This applies to both men and women.

- Secretary of the Navy Ray Mabus
- Bob Johnson, undersecretary of the Navy for Acquisition, Technology, and Logistics, meets with Phil Dert, a construction worker.

close proximity—Do not use; it's redundant. All proximity is close.

coalition—Do not capitalize.

- U.S. and *coalition* forces took part in the event.

Coast Guardsman—Capitalize in all references to U.S. Coast Guard.

- Sailors and *Coast Guardsmen* are instrumental in patrolling the Caribbean for drug smugglers.
- The local *coast guardsmen* work with Sailors to protect harbors.

Commander in Chief—Use only for the president. Always capitalize. Do not hyphenate.

commanding officer—Do not capitalize except when directly proceeding the title and name.

- *Commanding Officer* CAPT Tom Jones welcomed the distinguished visitors to the base.
- The *commanding officer* of the cruiser, CAPT Mary Smith, announced the ship would make a port visit to Key West, Fla.

CONUS—Continental United States. *CONUS* refers to the 48 contiguous states. It is not synonymous with United States. Do not use unless in a quote.

crew member—Use as two words. Do not use "crewman" or "crewmen." See **service members**.

currently—Avoid use. This term is redundant by nature. The ship is underway. Not—The ship is *currently* underway.

datelines/cities—For cities that stand alone, use the list of datelines found in *AP Stylebook*. Because of their strong Navy ties, Norfolk, San Diego, and Pearl Harbor can stand alone, without states.

dates—Follow the guidelines in the *AP Stylebook*.

D-Day—*D-Day* was June 6, 1944, the day the Allies invaded Europe during World War II.

decommissioned ships/submarines—Include reference that ship or submarine is no longer active.

- The *decommissioned* aircraft carrier *Constellation* (CV 64) will serve as a museum.

departments—Do not capitalize.

- The USS *Carl Vinson* engineering department
- The engineering department aboard the aircraft carrier

dependent—Do not use when referring to family of military personnel. Use terms such as "family members," "wife," "husband," "spouse," "parent," "child," etc. *Dependent* is perceived as derogatory. Do not identify *dependents* by name in photo captions.

detachment—Abbreviate as "Det." in all references.

- Helicopter Anti-Submarine Squadron Light (HSL) 43, *Det.* 5 also participated in the exercise.

DEVRON—submarine development squadron

doctor—Navy *doctors* are identified as "CMDR John W. Smith, a Navy *doctor*" in the first reference and by last name thereafter. See **military titles**.

DOD/DoD—Department of Defense. *DOD* or Pentagon is acceptable on second reference.

dry dock, **dry-dock**—Two words as a noun and hyphenate as a verb. Do not use as one word (see *Webster's*).

E-1 through E-3 Sailors—The term refers to enlisted Navy members in pay grades E-1 to E-3. These Sailors are identified as seaman recruit (SR), seaman apprentice (SA), or seaman (SN). Capitalize when directly preceding a name. The community variations of this naming convention are airman, constructionman, fireman, hospitalman, or seaman.

effect, **affect**—"Effect" is to cause, "affect" is to produce an effect upon (see *Webster's*).

ensure, **insure**—"Ensure" is a guarantee, while "insure" means to put insurance on something.

exercises—Use full title on first reference. Omit the word "exercise" on second reference: *Exercise* Kernal Potlatch '16, Operation Imminent Thunder. On second reference use Kernal Potlatch or Imminent Thunder. If *exercise* is abbreviated, follow the rules under **abbreviations and acronyms**.

- Example: RIMPAC '18

fast-attack—Hyphenate when used as an adjective.

- The *fast-attack* submarine deployed in November.

fleet—Use numerals and capitalize when referring to a specific *fleet* (U.S. 6th *Fleet*, U.S. 2nd *Fleet*, U.S. 7th *Fleet*). Do not capitalize in common usage: We sent a message to the *fleet*.

fleetwide—Use as one word.

flight deck—Use as two words.

fo'c'sle—A superstructure at or immediately aft of the bow of a vessel, used as a shelter for stores, machinery, etc., or as quarters for Sailors. It can also be written as "forecastle."

foreign cities—On first reference, the name of *foreign cities* are followed by the spelled-out name of the nation in which the city is located (e.g., Worms, Germany) unless listed in *AP Stylebook* under datelines.

frontline/front line—Use either as a noun or as an adjective.
- Troops on the *frontline* need supplies.
- *Front line* troops are the most in need.

general quarters—Lowercase when spelled out: The crew stayed at *general quarters* for 18 hours. Do not use "GQ."

global war on terrorism—Do not capitalize.

gray—Not "grey," except "greyhound."

guided-missile—Hyphenate when used as an adjective.
- The *guided-missile* cruiser is homeported in San Diego.
- The *guided missile* is loaded into a launch tube.

half-mast, half-staff—On ships and at naval stations ashore, flags are flown at *half-mast*. Elsewhere ashore, flags are flown at *half-staff*.

hangar, hanger—A *hangar* is a building; a *hanger* is used for clothing.

HCC—helicopter control center

HCS—helicopter combat support squadron

helo—A short, acceptable slang form of the word "helicopter."

homeport—One word in all uses.

- The Navy's newest *homeport* will be Detroit.
- The ship is *homeported* in San Diego.

HS—helicopter antisubmarine squadron

HSL—helicopter antisubmarine squadron light

hull number—See **ship names**.

Humvee—A trademarked abbreviation used for High-Mobility Multipurpose Wheeled Vehicle (HMMWV).

in country/in-country—Service members arrive *in country*. Once there, they have an *in-country* presence.

in port—Use as two words.

knot—A "knot" is one nautical mile (6,076.10 feet) per hour. It's redundant to say "*knots* per hour." Always use figures.

- Winds were at 7–9 *knots*; a 10-*knot* wind.

liaison—"Liaison" is a noun; do not use the verb form "liaise," as it is not usually used appropriately or well.

lifestyle—Use as one word.

Littoral combat ship—Do not capitalize.

- The *littoral combat ship* USS *Freedom* (LCS 1) arrived yesterday.

Marines—This is a proper noun. Capitalize when referring to U.S. forces (the U.S. *Marines*, the *Marine* Corps). Do not use the abbreviation "USMC."

maritime security operations—Lowercase when spelled out.

maritime strategy—lowercase

Mark—Do not use "MK" when referring to the word "Mark" in weapons or equipment.

- He worked on a *Mark* 50 torpedo.

master chief petty officer—Refers to Navy or Coast Guard personnel in pay grade E-9.

Master Chief Petty Officer of the Navy—Lowercase when referenced after an individual's name or when used alone. "MCPON" is the accepted abbreviation on second reference.

men—Do not use "men" if referring to a group of persons made up of *men* and women or whose genders are unknown. Use "Sailors" or "Marines" if the group is naval.

MIA—Missing in Action. *MIA* is acceptable on first reference.

midshipman—On first reference: *Midshipman* 1st Class John P. Jones (or 2nd Class, 3rd Class, 4th Class, if known). On subsequent reference(s): Jones.

- Note that the military abbreviation is MIDN 1/C (or 2/C, 3/C, 4/C, respectively): 1/C are in their senior year of school (USNA or NROTC), 2/C are juniors, 3/C are sophomores, 4/C are freshmen. "Midshipman" is singular, while "midshipmen" is plural. The term applies to both male and female.

military rank—The first reference should include rank and first and last name. All subsequent references should be last name only.

- Always refer to Sailors by rank and not pay grade (e.g., Yeoman 2nd Class or PO2, not E-5; CAPT or captain, not O-6). Follow the guidelines for spelling each rank in the *AP Stylebook*.

military titles/job titles—For enlisted personnel, spell out the Sailor's rate when directly preceding a name. Use petty officer(s)/airman, constructionman, fireman, hospitalman, or seaman when generalizing.

military units—Use numerals for unit designations. See **aircraft**, **fleet**, and **ship names**.

millimeter—Abbreviate as "mm" with no space or hyphen: 35mm film, 105mm howitzer, 9mm service pistol.

minehunter—Use as one word.

mine hunting—Use as two words.

missiles—Capitalize the proper name, but not the word "missile": Titan II *missile*.

naval—lowercase

naval activities—Spell out on first reference and capitalize only when part of a proper name: Naval Station Rota, Spain. On second reference, abbreviate as follows:
- naval station: NAVSTA
- naval air station: NAS
- naval weapons station: NWS
- naval amphibious base: NAB
- naval air facility: NAF
- construction battalion center: CBC

Navy Reserve—Capitalize when referring to the specific organization. Capitalize "Reserve" when referencing the U.S. Navy Reserve.

Navywide—Use as one word and always capitalize.

numbered fleets—Always refer to as digits and precede with "U.S."
- The ship is assigned to U.S. *6th Fleet*.

OEF, OIR—Operation Enduring Freedom, Operation Inherent Resolve. Do not add any number to these titles based on rotation status.

officer in charge—Do not hyphenate.

offload/off-load—One word as a noun and hyphenate as a verb. Do not use "upload" or "download."
- The *offload* took 12 hours.
- Deck department prepared to *off-load* pallets.

pay grade—Use as two words. Designations such as O-3, E-6, etc., are used only in reference to pay grades.

pendant—A short line-and-hooking device used to secure large objects (e.g., a cargo pallet or boat) to a towing or hoisting line.

pennant—A long, tapering flag used for signaling and/or identification.

percentages—Use figures.
- The crew's donation to Toys for Tots increased *20 percent*.

Persian Gulf—Use Arabian Gulf. "Gulf" is acceptable in second reference. Note that the Arabian Sea is its own body of water and should not be confused with references to the *Arabian Gulf*.

petty officer—Applies to Sailors or Coast Guardsmen in pay grades E-4 to E-6.

pierside—Use as one word.

plankowner—Use as one word.

Pre-Commissioning Unit (PCU)—Use for a ship not yet commissioned.
- The littoral combat ship *Pre-Commissioning Unit Little Rock* is undergoing acceptance trials.

pre-positioned—Hyphenate when referencing equipment or ships placed somewhere at an earlier date.
- Supplies and equipment were *pre-positioned* in the Gulf.

punctuation—When typing copy, leave only one space after all forms of punctuation. Follow guidelines in *AP Stylebook*.

quotation marks—The period and comma always go within the quotation marks. The dash, semicolon, question mark, and exclamation point go within the quotation marks when they apply to the quoted matter only. They go outside when they apply to the whole sentence.
- USS *Defender* (MCM 2) was awarded the Battle "E."
- "The crew performed superbly," said LCDR Charlie Brown.

rate—Refers to enlisted pay grades, e.g., E-4, E-8. Spell out and do not use warfare designations or qualifications.

reenlist—Do not hyphenate between the double vowel.

refueling complex overhaul (RCOH)—Lowercase when spelled out; uppercase the acronym.

replenishment-at-sea—Lowercase and use as a hyphenate.

Reserve, Reservist(s)—Capitalize "Reserve" when referencing the U.S. Navy Reserve.

retired—Use "retired" before rank/rate and name. Do not capitalize. Do not use abbreviated after a name.
- They invited *retired* Boatswain's Mate 1st Class John D. Writer.

Sailor—"Sailor" is to be capitalized in all references except when referring to seamen who belong to foreign navies or in merchant/civilian references.

Sea Hawk—Two words as the name of the FGA.6 aircraft. If referring to the SH-60 helicopter, write as one word, "Seahawk."

SEAL—Sea, Air, Land. "SEAL" is acceptable on first reference. If plural, use "SEALs."

Secretary of the Navy (SECNAV)—Lowercase when referenced after an individual's name or when used alone. SECNAV is the accepted abbreviation on second reference.

service member(s)—Use as two words.

ship names—For first reference always include "USS," the ship's name, and the hull number: USS *Harry S. Truman* (CVN 75).
- Do not use "USS" for ships before 1909, or if it is not yet in commission, or has been decommissioned and you are referring to the ship in its present state.
- There is no hyphen in the hull number. On second reference use only the ship's name or reference as "the ship." Do not use "the" in front of a ship's

name: "USS *San Jose*," not "the USS *San Jose*." Use "the" before a ship type: "the aircraft carrier USS *George H. W. Bush* (CVN 77).

- Ship names are in italics, not all caps. Use USS *Seattle*, not USS SEATTLE.

Ships' nicknames—Do not use. "The aircraft carrier USS *Dwight D. Eisenhower* (CVN 69) is underway. *Dwight D. Eisenhower* will deploy to the U.S. 5th Fleet." In official writing use only official names of ships, aircraft, or people.

spokesperson—"Spokesman" or "spokeswoman" is preferred. Use "spokesperson" only if the gender of the individual is unknown. If possible, use a generic term instead: "public affairs officer (PAO)," "representative," etc.

squadron—Spell out on first reference and use numerals for the squadron's number: Fleet Air Reconnaissance *Squadron* (VQ) 1. On second reference, use the appropriate abbreviation (with a hyphen), e.g., VQ-1.

stand down/standdown—One word when used as a noun. Two words when used as a verb.

- The safety *standdown* was held Nov. 4.
- The officer in charge told him to *stand down*.

state names—See *AP Stylebook* for proper usage.

Submarine Force—Use upper case when referring to this section of the fleet.

SUBRON—submarine squadron

team member—Use as two words.

theater security cooperation (TSC)—Lowercase when spelled out; uppercase acronym.

time—Do not use military time unless quoted.

titles—Capitalize titles when used before a name only. See "titles" entry in *AP Stylebook*.

undersecretary—Use as one word.

underway—Use as one word.

VAQ—electronic-attack squadron

VAW—carrier airborne early warning squadron

VFA—strike-fighter squadron

VMFA—Marine strike-fighter attack squadron

VP—patrol squadron

VRC—fleet logistics-support squadron

VS—sea-control squadron

VT—training squadron

warfare qualifications and designations—Do not include an individual's warfare qualification after their name, as in "PO1 Jones, ESWS."

warfighter—Use as one word.

washdown—Use as one word.

watchstander, watchstanding—Use as one word.

wide—Close up this suffix with most words: "nationwide," "Navywide," "fleetwide," etc.

woman, women—Preferred to "female."

Index

About the Authors

Robert Shenk (1943–2021) was a widely published professor of English at the University of New Orleans and a retired captain in the U.S. Navy Reserve. He served on a destroyer and on river patrol boats during the Vietnam War and later taught at two service academies. He lived in Mandeville, Louisiana.

Dr. Christopher "Chip" Crane, USNA Class of 1990, has been teaching and training others in writing for more than twenty-five years. He consults for the federal government and professional organizations in writing, presenting, leadership, team building, and strategic planning. Dr. Crane also teaches professional and technical writing at the University of Maryland at College Park. His twenty years in the Navy included a tour on USS *Theodore Roosevelt* (CVN 71), a deployment to Afghanistan, and retail and business management billets. From 1999 to 2010, he also served on the faculty of the Naval Academy's English department, teaching professional writing and literature and directing the USNA Writing Center.

THE NAVAL INSTITUTE PRESS is the book-publishing arm of the U.S. Naval Institute, a private, nonprofit, membership society for sea service professionals and others who share an interest in naval and maritime affairs. Established in 1873 at the U.S. Naval Academy in Annapolis, Maryland, where its offices remain today, the Naval Institute has members worldwide.

Members of the Naval Institute support the education programs of the society and receive the influential monthly magazine *Proceedings* or the colorful bimonthly magazine *Naval History* and discounts on fine nautical prints and on ship and aircraft photos. They also have access to the transcripts of the Institute's Oral History Program and get discounted admission to any of the Institute-sponsored seminars offered around the country.

The Naval Institute's book-publishing program, begun in 1898 with basic guides to naval practices, has broadened its scope to include books of more general interest. Now the Naval Institute Press publishes about seventy titles each year, ranging from how-to books on boating and navigation to battle histories, biographies, ship and aircraft guides, and novels. Institute members receive significant discounts on the Press' more than eight hundred books in print.

Full-time students are eligible for special half-price membership rates. Life memberships are also available.

For more information about Naval Institute Press books that are currently available, visit www.usni.org/press/books. To learn about joining the U.S. Naval Institute, please write to:

Member Services
U.S. NAVAL INSTITUTE
291 Wood Road
Annapolis, MD 21402-5034
Telephone: (800) 233-8764
Fax: (410) 571-1703
Web address: www.usni.org